Crystals in Gels and Liesegang Rings

IN VITRO VERITAS

Crystals in Gels and Liesegang Rings

HEINZ K. HENISCH
The Pennsylvania State University

The right of the
University of Cambridge
to print and sell
all manner of books
was granted by
Henry VIII in 1534.
The University has printed
and published continuously
since 1584.

CAMBRIDGE UNIVERSITY PRESS

Cambridge

New York   New Rochelle   Melbourne   Sydney

Published by the Press Syndicate of the University of Cambridge
The Pitt Building, Trumpington Street, Cambridge CB2 1BP
32 East 57th Street, New York, NY 10022, USA
10 Stamford Road, Oakleigh, Melbourne 3166, Australia

First published 1988

Printed in Great Britain by J. W. Arrowsmith Ltd, Bristol.

British Library cataloguing in publication data

Henisch, Heinz K.
Crystals in Gels and Liesegang Rings.
1. Crystals – Growth
I. Title
548′.5   QD931

Library of Congress cataloguing in publication data

Henisch, Heinz K.
Crystals in Gels and Liesegang Rings: in vitro veritas/Heinz K. Henisch.
p.   cm.
Includes index.
ISBN 0 521 34503 0
1. Crystals – Growth.   2. Colloids.   3. Liesegang Rings.
I. Title
QD921.H473 1988
548′.5 – dc19   87-17203   CIP

ISBN 0 521 34503 0

To Bridget

CONTENTS

PREFACE

When *Crystal Growth in Gels* (The Pennsylvania State University Press, University Park, Pennsylvania, 1970) was published, it was very much a playful 'first book' on the subject. Since then, the field as a whole has flourished, both in terms of practical technique and in terms of understanding. Inevitably, the advent of the computer has also left its mark. The present volume was prepared in order to provide an updated summary of our experience. *Inter alia*, it is intended as a guide to the literature, but it should not be taken as an encyclopedic or exhaustive evaluation of previously published material.

I am indebted to the late Dr. V. Vand for introducing me to the subject of crystal growth in gels in the 1960s and to many friends for their collaboration, their advice and, above all, for their forebearance in the face of enthusiasm. On this occasion, I would like to express special thanks to colleagues who have so generously contributed new illustrative material to this edition: J. Adair, H. Arend, S. K. Arora, H. Behm, A. S. Bhalla, B. Březina, J. F. Charnell, J. M. García-Ruiz, M. T. George, E. S. Halberstadt, J. Hanoka, F. Lefaucheux, P. Ramasamy and J. Ross. I owe a special debt to J. M. García-Ruiz for many stimulating discussions, to Bonny Farmer for the exercise of her editorial skills, and to my wife, Bridget, for soothing the furrowed brow in times of crisis.

University Park, PA   H.K.H.
August 1986

LIST OF SYMBOLS

(A)	identification of reagent (A)	First used in section 2.3
$A(X, T)$	concentration of (A) as a function of X and T	2.3
A_G	initial concentration of (A) in the gel (uniform)	5.4
A_R	reservoir concentration of (A)	2.3
A_s	saturation value of A	3.1
$A_{\min}$	minimum value of A for precipitation	5.5
A_N	value of A for $X = N\,\Delta X$, with $\Delta X = 1$	5.6
$\mathscr{A}$	cross-sectional area	5.4
(B)	identification of reagent (B)	2.3
$B(X, T)$	concentration of (B) as a function of X and T	2.3
B_G	initial concentration of (B) in the gel (uniform)	5.4
B_R	reservoir concentration of (B)	2.3
$B_{\min}$	minimum value of B for precipitation	5.5
b	constant	5.5
C	concentration in general	3.1
C_∞	concentration at which 'particles of infinite radius' would nucleate	4.1
C_r	concentration at which particles of radius r nucleate	4.1
D	diffusion coefficient in general	5.4
D_A, D_B	diffusion coefficients of reagents (A) and (B)	2.3
E_c	critical energy for nucleation	4.1
$F(s)$	Frank's function, relating concentration gradients and growth rates in terms of s	3.1
G	constant	5.3
G_F	growth factor	5.6

H	heat of solution	4.1
h	constant, spacing coefficient	5.3
i	counter	5.3
K_s	concentration product for saturation	3.6
K_s'	concentration product for precipitation	3.6
K_s''	a concentration product $> K_s'$	5.5
k	Boltzmann's constant	4.1
L	total length of a diffusion system in units of ΔX	2.3
M	molecular weight	4.1
$\mathcal{M}_A, \mathcal{M}_B$	masses of reagents (A) and (B) crossing boundary in unit time	5.4
m	crystal mass	4.3
N	$X_N = N \, \Delta X$, usually with $\Delta X = 1$	2.3
n	concentration of seeds	4.3
P_N	nucleation probability	5.5
p_r	vapor pressure immediately outside a droplet of radius r	4.1
p_∞	vapor pressure immediately outside a flat liquid surface	4.1
R	radial distance from a growing crystal	3.1
R_B	the Boltzmann gas constant	4.1
R_{DC}	reservoir depletion coefficient	5.6
r	crystal 'radius'	3.1
r_c	critical radius	3.1
S	surface area	4.3
s	'reduced radius' $= r/(DT)^{1/2}$	3.1
T	time	2.3
T_N	time of formation of the Nth Liesegang Ring	5.6
t	temperature	
v_A, v_B	partial molar volumes of the (A) and (B) species	3.1
X, x	distance	2.3
X_N	position of the Nth Liesegang Ring	5.6
Y, y	distance alternatives to x and X	3.1
Y_i	position of the ith Liesegang Ring	5.3
z	integration variable	5.4

1

History and nature of the gel method

1.1   Introduction

It has long been appreciated that advances in solid state science depend critically on the availability of single crystal specimens. As a result, an enormous amount of labor and care has been lavished on the development of growth techniques. In terms of crystal size, purity, and perfection, the achievements of the modern crystal grower are remarkable indeed, and vast sections of industry now depend on his products. So do the research workers whose preoccupation is with new materials, no matter whether these are under investigation for practical reasons or because a knowledge of their properties might throw new light on our understanding of solids in general.

In one way or another, a very large number of materials has already been grown as single crystals, some with relative ease and others only after long and painstaking research. Nevertheless, there are still many substances which have defied the whole array of modern techniques and which, accordingly, have never been seen in single crystal form. Others, though grown by conventional methods, have never been obtained in the required size or degree of perfection. All these constitute a challenge and an opportunity, not only for the professional crystal grower but, as it happens, also for the talented amateur. New and unusual methods of growing crystals are therefore of wide interest; and if the crystals are by themselves beautiful, as they often are, there is no reason why this interest should be confined to professional scientists.

The art and science of growing crystals in gels enjoyed a long period of vogue close to the end of the last century and lasting well into the 1920s. It remained largely dormant during the 1930–60 period and then experienced a mini-renaissance which held great promise for the future. It would now be pleasing to be able to report on rapid and sustained progress ever since, but for a variety of reasons, developments have in fact been slow, especially as far as the more fundamental implications of the growth process are concerned. On the positive side, the astute reader will note that this leaves the research opportunities as great as and as open as they have ever been.

During most of the early period, the center of interest was held by the phenomenon of Liesegang Rings. Liesegang was a colloid chemist, a photographer, and an altogether remarkable man (see Section 5.1), who experimented with chemical reactions in gels (e.g. see Liesegang, 1896, 1898, 1924, 1926).† He covered a glass plate with a layer of gelatin impregnated with potassium chromate, and added a small drop of silver nitrate. As a result, silver chromate was precipitated in the form of a series of concentric rings, well developed and with regularly varying spacings. Such structures had actually been reported in the earlier literature, but Liesegang was the first to take them seriously. Runge (1855) had observed them in the course of experiments on the precipitation of reagents in blotting paper! Towards the end of the nineteenth century interest was regenerated in the context of photographic gelatin emulsions (e.g. see Lüppo-Cramer, 1912), and the discontinuous nature of the precipitation, its geometrical features, and the conditions of its occurrence, soon became objects of intense, if not altogether successful, investigations. The matter immediately attracted the attention of the great German chemist Ostwald (1897a, b) and, in due course, need one mention it, that of Lord Rayleigh (1919), thereby receiving what must have appeared as the ultimate seal of respectability. According to a report by Bradford (1926), J. J. Thomson, whose principal interests were in very different fields, likewise concerned himself with the problem of periodic precipitation (but the Royal Institution appears to have no record of the 1912 lecture to which Bradford refers). Oddly enough, all the intense turn-of-the-century activity in the field did not prevent Janek (1923) from reporting periodic precipitation as a new phenomenon. See Deiss (1939) for an extensive overview of phenomena leading to 'Runge Bilder', some reproduced in color.

Liesegang's structures were (quite properly) called rings, because they were first observed in that form, i.e. as concentric deposits in a plane. Later they were more often grown in test tubes, and were therefore disks, but the 'ring' designation was by then firmly implanted in the literature, and has prevailed to this day. Rings were considered interesting partly

† See bibliography at the end of the book.

because their origin was obscure and partly because they were reminiscent of certain structures found in nature, e.g. the striations of agate. Often they consisted of apparently amorphous material but, in due course, the achievement of microcrystalline reaction products came to be seen as desirable, because of the ease with which they could be identified by means of x-ray photographs. Larger crystals, several millimetres in size, were occasionally obtained, but not systematically looked for. In contrast, the growth of such crystals is the principal objective of most modern work in this field. This endeavor has, over the years, succeeded to a remarkable extent, but not (possibly not yet) sufficiently to make crystal growth in gels an industrial process. It remains a laboratory process, each specific substance grown being associated with problems of its own. An early claim by Fisher and Simons (1926a) to the effect that 'gels form excellent media for the growth of crystals of almost any substance under absolutely controllable conditions' survives as a shining example of faith, but is, even so, no more than a little tarnished by the sporadic nature of its fulfilment to date.

Surprisingly, in view of the history of the subject, the Liesegang Ring phenomenon itself is even now only partially understood. It has largely been displaced from the center of practical interests, but has in recent years achieved a new intellectual status in the context of fundamental 'order out of chaos' discussions (shades of Prigogine, 1984). It also awaits the talents of a modern Ostwald, with an appreciation of its beauties and more versatile resources than Ostwald was ever able to command. There is no basic mystery, in the sense that periodic solutions of the diffusion equations (with proper allowance for precipitation, resolution and boundary conditions) are known to exist, but little is known about the detailed parameters involved. Even so, a substantial beginning has been made, and recent work, partly with the aid of the computer, makes it possible to relate theory to practice in a general and consistent way. Chapter 5 of the present volume is devoted specifically to Liesegang Ring problems.

Of course, periodic ring and layer formations *found in nature* offer only the most limited opportunities for research into their origin. Indeed, many are due to quite different mechanisms, e.g. changes of overall environment, even though they bear the superficial appearance of Liesegang Rings. As one critical analyst has put it when faced with the suggestion that the stripes of tigers and zebras may be glorified Liesegang phenomena: 'enthusiasm has been carried beyond the bounds of prudence' (Hedges, 1931), a verdict with which the present writer is inclined to concur. In the same spirit of caution, we may want to reject the notion that Liesegang phenomena are responsible for the stripes on butterfly wings (Gebhardt, 1912).

Though geology-at-large does not permit experimentation, work during the early period derived a good deal of impetus from the interests of geologists, who believed that all quartz on earth was at one time a silica hydrogel. A vein of gelatinous silica, as yet unhardened by dehydration, was indeed reported to have been found in the course of deep excavations for the Simplon tunnel (Spezie, 1899). Von Hahn (1925) described a similar find on the Lüneburger Heide. (He asked a cardinal question, designed to quicken our interest: Can you eat it?, and remarked that the substance had a 'cool, refreshing, slightly astringent taste', and a smell reminiscent of fruit bonbons!) Moreover, some early experiments were on record, quoted by Eitel (1954), according to which microscopic silica crystals had been obtained from silica gels in the presence of various 'crystallizing agents' when heated in water vapor under pressure. Despite the exotic character of some of these findings, it is entirely plausible that crystalline foreign deposits which are so often found in quartz may be examples of crystal growth in gel. In this way, the gel method appeared to offer systems and opportunities for experiments in 'instant geology' (e.g. see Holmes, 1917), and indeed it does so.

Fig. 1.1.1 shows typical examples of natural growth, needles of tourmaline and rutile in a single crystal of quartz. The needles must have grown first, and it is tempting to believe that they did so when the medium was a gel, though current opinion among geologists and mineralogists leaves room for other interpretations, e.g. that the growth may have been hydrothermal in character, with convection playing a crucial role. On the other hand, little is known about the viscosity of the hydrothermal medium under the original growth conditions, and it may well be that the tourmaline and rutile crystals grew when the fluid was viscous. In such a case, the distinction between hydrothermal growth and growth in a gel would be blurred. This view is supported by the knowledge that a gel as such, though generally beneficial, is not absolutely required for the mode of crystal growth under discussion here (see below). On the role of gels in geology, there is an extensive early literature (e.g. Cornu, 1909 and Krusch, 1907, 1910). In the last analysis, the problem of consistency has not yet been unequivocally settled as far as quartz is concerned, but there are many examples of crystal growth in other viscous media, natural (Koide and Nakamura, 1943) and artificial. Among the unwelcome manifestations of the process are the occasional growth of ice crystals in ice cream, the growth of tartrate crystals in cheese, the crystallization of sulfur in rubber (Endres, 1926), the growth of zinc salts in dry batteries and, in rapidly descending order of desirability, the growth of uric acid crystals in joints (Knöll, 1938a, b) and of stones in human organs (Aschoff, 1939). The subject thus has much wider implications than is generally believed. Thus it is also related to our understanding of the processes which take place in photographic emulsions.

Fig. 1.1.1. Natural growths: needles of (*a*) tourmaline, and (*b*) rutile, in quartz.

(*a*)

(*b*)

1.2   Early work

In the course of early work a mass of empirical data was assembled, much of it too imprecise and unsystematic to lead to any real insight into the mechanism of the phenomena involved. However, some of these investigations remain interesting because they have the character of 'existence theorems', illustrating at least some of the things that can be done and suggesting new lines of approach.

Among the indefatigable enthusiasts was Hatschek (1911), working primarily with (5-20%) gelatin and (1-5%) agar gels. He was the first to make a study of particle size distribution in a great variety of Liesegang Rings. Among other things, he also noted that even when good rings are formed, they are not necessarily the only reaction product. When a ring

Fig. 1.2.1. Liesegang Rings: (*a*) silver chromate system. (*b*) calcium phosphate system.

(*a*)   (*b*)

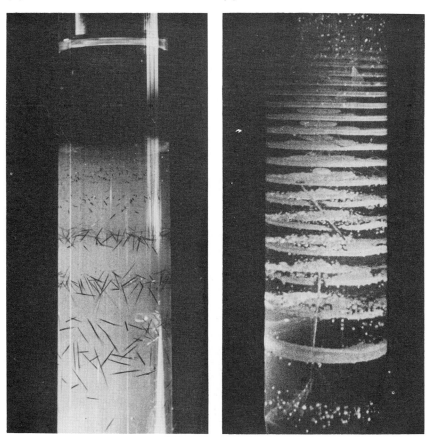

system is prepared by allowing sodium carbonate to diffuse in a gel charged with barium chloride, distinct and well-formed barium carbonate crystals up to 1.5 mm in length are occasionally found in the otherwise clear space between adjacent rings. The observations must have been the first hint, if one was needed, that Liesegang Ring information is indeed a complicated process, especially since Ostwald and Morse and Pierce (1903) had earlier placed great emphasis on the sharpness with which the rings are often defined. The work of these researchers will be more extensively discussed in other parts of this book. In any event, it has since been amply demonstrated that the rings themselves may consist of substantial crystals (Fig. 1.2.1(*a*)).

Silver chromate, and the dichromate, chromate, chloride, iodide, and sulfate of lead, as well as calcium sulfate and barium silicofluoride

Fig. 1.2.1 (continued). Liesegang Rings: (*c*) gold system.

(*c*)

were included in Hatschek's crystal-growing repertoire. Hatschek was also the first to point out that crystals often grow better in silicic acid gels than in gelatin or agar, but it is not actually possible to generalize such a rule; there is no such thing as a universal 'best' growth medium. In one of the most truly interdisciplinary experiments on record, Marriage (1891) had earlier grown lead iodide crystals in fruit jellies and jams.

A description of the basic procedures will be found in the next section. It was always realized that agents could be diffused into a gel, and conversely, that gels could be dialyzed to free them from excess reagents or unwanted reaction products. Such experiments were conducted by Holmes (1926), who used the dialyzing process for the treatment of gels in U-tubes, in order to eliminate excess reagents that could have interfered with the two principal components of the formation process diffusing in from the two sides. Holmes also grew single crystals of copper and gold by diffusing a reducing agent into a gel charged with appropriate metal salts. Hatschek had done so before, but Holmes claimed 'better results than any yet recorded', without actually mentioning crystal sizes. Both experimenters concerned themselves to some extent with the effect of non-reagent additives, such as glucose, urea, and gum tragacanth. Holmes and also Davies (1923) noted that reactions in gels can occasionally be influenced by light and, in particular, by short wavelength, ultraviolet radiation; see Section 5.2. Holmes used alkaline gels to promote the formation of cuprous oxide layers, but though he aimed at cuprous oxide crystals (perhaps because of their interesting electrical and photoelectric properties) the product was always amorphous.

One of the most important early experiments was performed by Dreaper (1913), who wanted to elucidate the role played by the capillarity of the gel structure. For this purpose, he substituted fine sand and even a single capillary tube for the usual gels and found that crystalline growth products would indeed be obtained with such systems. Many crystals do, in fact, grow under such conditions in nature, e.g. ice in soil and pure gypsum crystals in clay. Holmes later used barium sulfate and alundum powders for the same type of demonstration. Of course, these experiments prove nothing in particular about the nature of gels, but they provide a valuable hint of the circumstances which favor single crystal formation; see Section 4.4.

A comprehensive survey of early work on gel structures has been given by Lloyd (1926). Structure and classification problems were hotly debated by scientific workers in the 1920s and earlier, often with enviable self-confidence and just occasionally with a trace of venom (e.g. see von Weimarn, 1926). During the next 30 years or so, the subject of crystal growth in gels fell into virtual oblivion, until it was revived by the modern interest in outlandish materials and, more generally, in room-temperature growth methods. In comparison, an enormously higher level of interest was sustained throughout this period in Liesegang Ring phenomena, a fact for which bibliographies by Hedges (1932) and Stern (1967) comprising many hundreds of publications bear eloquent testimony. Some beautiful color photographs of Liesegang Ring formations have been published by Kurz (1965).

1.3   Basic growth procedures

To all outward appearances, the gel method is exceedingly simple, but it is now abundantly clear that the physical and chemical processes which determine its outcome are in fact highly complex. One procedure, much used for demonstration purposes (e.g. see Henisch *et al.*, 1965), involves the preparation of a hydrogel from commercial waterglass, adjusted to a specific gravity of about 1.06 g/cm. This is mixed with an equal volume of approximately 1 molar (1 M) acid solution. One's natural inclination is to pour the acid into the waterglass (or metasilicate, see below) solution, but that leads to disaster. Instead, the waterglass should be added drop by drop to the acid, with constant stirring. The purpose of this procedure is to avoid excessive ion concentrations which otherwise cause premature local gelling and make the final medium inhomogeneous; see also Madeley and Sing (1962). The mixture is then allowed to gel, a process which depends in a complicated manner on the silicate concentration and on the degree of acidity (see below and also Chapter 2). In the circumstances specified, it generally takes between 24 and 36 hours, after which the gel is left for another 12 hours or so to allow it to set firmly. Once a firm gel is formed, some other solution can be placed on top of it without damaging its surface, as shown in Fig. 1.3.1(*a*). This solution supplies one of the components of the reaction and also prevents the gel from drying out. In order to avoid damage, the supernatant solution should be added dropwise with a pipette, the drops being allowed to fall on to the side of the test tube. If the reagent in the gel is tartaric acid and the supernatant reagent is an approximately 1 M solution of calcium chloride, then, in due course, crystals of calcium tartrate tetrahydrate are formed in the gel, the result of an exchange reaction, which yields hydrochloric acid as a by-product. The crystals are only sparingly soluble in water. They can usually be seen near the gel surface within an hour or so. Good crystals appear further down the column within about a week, and grow to about 8 mm average size. Occasionally they are larger, but then usually less perfect. The speed of formation depends on the concentrations involved, and the time taken before crystals are seen with the naked eye may vary from very few hours to a few days. When much higher acid concentrations are used, sodium tartrate crystals also tend to grow, in the form of long clear needles. Calcium tartrate crystals are beautiful, and the fact that they appear to be totally useless has always been a matter of profound regret within the crystal-growing community. Along these lines a ray of hope has recently appeared, in the form of a reported anomaly of the electrical conductivity (Gon, 1985), but these observations remain to be consolidated. There

Fig. 1.3.1. Basic growth procedures. During gel preparation the waterglass or sodium metasilicate solution should be added drop by drop to the acid with constant stirring. (*a*), (*b*) Simple test tube systems; (*c*) test tube system with tubular insert for easy gel removal; (*d*) U-tube system.

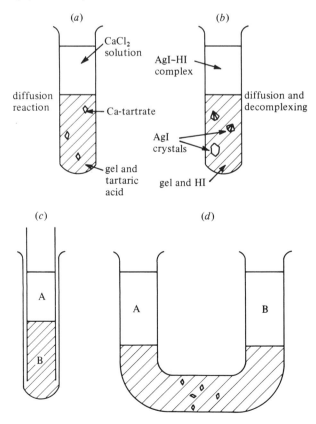

are even reports (albeit unconfirmed) that the material may be ferroelectric (Arend, 1986).

Because commercial waterglass does not have a fixed and accurately known composition, and because it is generally contaminated with undesirable impurities, it is better to use reagent grade sodium metasilicate. A stock solution is prepared by adding 500 ml of water (distilled or demineralized) to 244 g of $Na_2SiO_3 \cdot 9H_2O$. As far as possible, this solution should be kept from contact with the atmosphere to avoid absorption of carbon dioxide (Halberstadt, 1967a, b). As one would expect, the most consistent results are obtained when the gels are kept under thermostated conditions during crystal growth, e.g. between room temperature and 45 °C, though a really high degree of temperature stability does not seem to be required. Crystals of remarkably high optical perfection can be grown (Fig. 1.3.2). A more detailed discussion of the factors which govern ultimate size and degree of perfection will be found in Sections 3.2 and 4.5. As a general rule, very dense gels produce poor crystals. On the other hand, gels of insufficient density take a long time to form and are mechanically unstable. A specific gravity of 1.02 g/cm appears to be the lower practical limit.

Fig. 1.3.2. Growth of calcium tartrate crystals. (*a*) Acceptable nucleation. (*b*) Excessive nucleation. (*c*) Gel-grown calcium tartrate crystals.

(*a*)   (*b*)   (*c*)

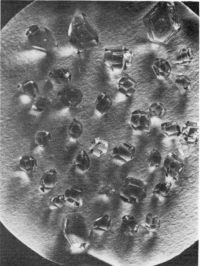

It is evidently a simple matter to vary the main parameters of this system. Thus, calcium tartrate crystals can be grown not only by using calcium chloride, but any other soluble calcium salt, such as calcium acetate. However, a straightforward comparison is difficult, because solutions of equal concentration differ in pH, and whereas the initial pH can be adjusted, the subsequent changes are not easily monitored and controlled.

A host of other crystals can be grown with varying degrees of success by using different acids and different salts. Ammonium, copper, cobalt, strontium, iron, and zinc tartrates, cadmium and silver oxalates, calcium tunstate, lead and silver iodides, mercuric iodide, calcium sulfate, calcite and aragonite, lead and manganese sulfides, metallic lead, copper and gold are among those which have been successfully prepared – and there are many others. Nor is it necessary for the second reagent to be in the form of a solution. Gas reagents under varying pressure can be used, and these offer the additional possibility of extending the temperature range within which the experiments are carried out. The gel itself need not necessarily be acidic (e.g. see Dennis *et al.*, 1965 and Fisher, 1928) nor need it be based on sodium metasilicate; various proprietary silicas ('Ludox', 'Cab-O-Sil') can also be used, as can agar gels, though with generally poorer results. The proprietary silicas are free (or relatively free) of sodium, which should be an advantage in principle, since sodium is usually a contaminant and not part of any essential reaction. On the other hand, gels with and without sodium, but of otherwise identical structures, have not yet been systematically compared. As a practical matter, the point therefore remains unsettled. In a series of early experiments on the growth of lead iodide and lead bromide, Fisher (1928) actually found the presence of sodium beneficial for large crystal growth, though it is hard to see why it should be so.

The formation of calcium tartrate crystals, as described above, is an example of a growth process involving a strong acid as the (unwanted) reaction product, and, of course, not all processes are of this kind. The growth of lead iodide, for instance, does not involve such acid formation. In a typical demonstration experiment, the stock solution (7.5 ml) of sodium metasilicate (see above) is diluted with an equal quantity of water. Then 15 ml of 2 M acetic acid and 6 ml of 1 M lead acetate are combined, with continued agitation, see Fig. 1.3.1 (caption). The mixture is allowed to set, and (say) 20 ml of 0.75 M potassium iodide are then placed on top of the gel in the manner described above. Good, but thin hexagonal plates of lead iodide, about 8 mm in diameter and occasionally larger, grow within about three weeks at 45 °C. Lower temperatures (e.g. 20 °C) favor greater thicknesses, at the expense of platelet diameter. Crystals up to 3 mm thickness have been grown by Dugan (1967); see also Section 5.1. When lead iodide and other growth systems are stored for long periods, various changes may take place due to the formation of double salts (Fisher and Simons, 1926a, b and Fisher, 1928).

Among other lead compounds so far prepared is lead sulfide which has been grown in acidic lead acetate gels, covered with a supernatant source of sulfur ions. Brenner and co-workers (1966) used dilute solutions (0.1 M) of thioacetamide for this purpose. This compound yields sulfur after reaction with the acid which diffuses out of the gel. Cubic crystals of over 1 mm size were obtained. Somewhat smaller crystals (0.5 mm edge) have been reported by Murphy and Bohandy (1967), who used sodium sulfide as the sulfur source. Section 3.6 deals with lead sulfide in greater detail.

Calcium tartrate and lead iodide are both formed in initially acidic gels, though the lead iodide system becomes increasingly alkaline in the course of growth. Lead hydroxy-iodide, $PbI(OH)$, is an example of a crystal which can (and, indeed, must) be grown in gels which are alkaline from the start. (See also Section 3.4 for notes on the growth of calcite and other carbonates.) In this case, the sodium metasilicate and acetic acid solution is adjusted to a pH of 8 by means of potassium iodide, and allowed to gel. The hope is that the unwanted potassium will not get in the way, and as far as is known it does not. Lead acetate is then diffused into the gel from a supernatant solution, while the growth system is kept at 40 °C or above (Dennis *et al.*, 1965). The crystals which result from this procedure are of the order of 1 mm^2 in cross-section and up to 5 mm long. They are clear, pale yellow in color, and easily distinguished from the more orange hexagonal platelets of lead iodide with which they co-exist in some growth systems (Fig. 1.3.3). See also Miller (1937a).

The processes described above can be influenced electrolytically by the application of electric fields within the gel and, in particular, it is known that such fields can influence the structure and distribution of Liesegang Rings (Happel *et al.*, 1929a, b, Kisch, 1929a, b, Dhar and Chatterji, 1925a, and Christomanos, 1950). Occasionally, advantages have been reported for single crystal growth (e.g. see Desai and Rai, 1981, 1984, and Pillai *et al.*, 1980, 1981) but not in a sufficiently systematic and consistent way to lead to new understanding. It is, indeed hard to see how there could be a 'systematic and consistent way', considering that each growing crystal, whether conductive or not, modifies the local field pattern. However this may be, George and Vaidyan (1981a, b,

1982a, b) used an electrolytic technique to produce small but very fine crystals of metallic silver. For estimates of ionic mobilities in gels, derived from electrolytic transport experiments, see Swyngedauw (1939), but it should be remembered that gels are not highly standardized media.

Mixed crystals can be grown by using mixed reagents, and crystals can generally be doped by the use of small amounts of additives, either in the gel itself or in the top reagent (see below and Section 3.4). By changing the top reagent from one metal salt solution to another, it is sometimes possible to form heterojunctions (of varying degrees of abruptness) between isomorphous substances. Very little has so far been done along these lines, but the method is full of potential for sustaining investigations of theoretical and practical importance.

In the procedures described thus far, the gel is used as the reaction medium, in which the desired material is chemically formed. In a variant of the method, pioneered by O'Connor and co-workers (1966)

Fig. 1.3.3. Growth of PbI$_2$ and PbI(OH) crystals. (a) PbI$_2$ simple growth.

(a)

Fig. 1.3.3. Growth of PbI_2 and $PbI(OH)$ crystals. (b) PbI_2 with concentration programming. (c) $PbI(OH)$.

(b)

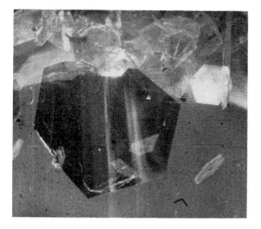

(c)

and Armington *et al.* (1967a, b) for the growth of cuprous chloride crystals, the material is first complexed by means of another reagent (hydrochloric acid), and then allowed to diffuse into a gel free of 'active' reagents. Decomplexing sets in with increasing dilution and leads to the high supersaturations necessary for crystal growth. Cuprous chloride is interesting as a possible laser modulator, in which modulation can be achieved by a transverse electrooptic effect (Murray, 1964). The crystals are ordinarily grown from the melt, but thermal strains arise as a consequence of cooling from the melting point of 422 °C. In addition, there is a phase change from the wurtzite to the zinc blende structure at 407 °C. This is actually the wanted form, but the cooling process is complicated by this transformation.

In view of these difficulties, the gel method has obvious advantages: the initial experiments described by Armington and co-workers (1967a, b) were carried out in 2 cm diameter test tubes containing pH 5 hydrochloric acid gels, with 5 ml of supernatant solution of varying acidity, saturated with cuprous chloride. Tetrahedral crystals of 3 mm size grew within about three weeks. As in the case of calcium tartrate, the gel tends to split off the growing crystal, especially at higher pH values, resulting in this case in the unwanted growth of numerous microcrystals. In the course of such test tube experiments, conditions in the gel become more acid with time, and crystals formed during the earlier diffusion process often redissolve later. This has also been reported by Kirov (1972), and is further discussed in Section 5.6.

Experiments with U-tubes (see Fig. 1.3.1(d)) have also been described, and better results are sometimes obtained by the use of this configuration, especially when large reagent reservoirs are also employed (see below). A useful variant of the U-tube configuration has been described by Patel and Arora (1973), as shown in Fig. 1.3.4(a) and (b). It permits the gradual removal of unwanted reaction products, and thus also the control of pH in the growth region. Clear crystals of barium tungstate ($BaWO_4$) and strontium tungstate ($SrWO_4$), both several millimetres in length, have been grown in that way, and exhibited a very low dislocation density; see also Lefaucheux *et al.* (1979), who used this technique for growing brushite crystals. Various elaborations of the method have been described by Patel and Rao (1979, 1980).

By and large, the best results have been obtained (Armington and O'Connor, 1968a) with systems which offered not only constant concentration reservoirs but also a constant path-length over the cross-section of the gel (Fig. 1.3.5). This also permitted the gel to be easily removed without breakage for recovery of the crystals of reagent analysis. Of all

Fig. 1.3.4. Double-tube growth systems with possibilities for pH and composition control in the growth region. (*a*) Single-tube version. (*b*) Double-tube version. (*c*) Barium tungstate crystals (grid in mm). (*d*) Strontium tungstate crystals (grid in mm). After Patel and Arora (1973); see also Arora (1981).

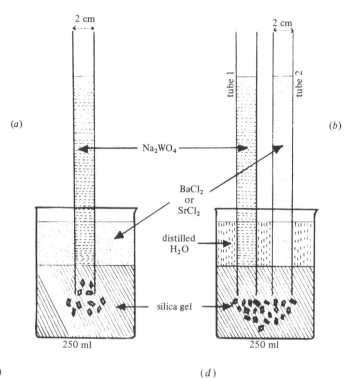

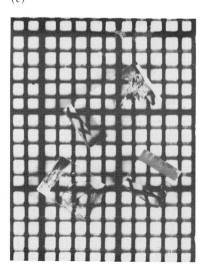

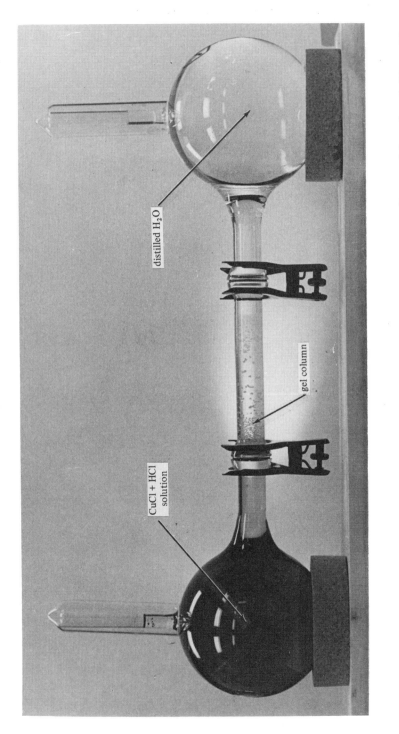

Fig. 1.3.5. Double-diffusion system with reagent reservoirs and straight diffusion column; diameter 2.0 cm, length 25 cm. After Armington and O'Connor (1968a, b).

the geometries, the linear column offers the best opportunity for predicting the optimum growth region from a knowledge of the solubility and the variation of pH with distance. Such estimates are simplest when the gel column is (for practical purposes) infinitely long, or else when it is of finite length but with well-defined boundary concentrations. Armington and O'Connor (1968b) have shown how such growth estimates can be obtained, as long as the crystals themselves are reasonably small compared with the diffusion cross-section. The use of linear columns and of large reagent reservoirs doubled the size of the crystals as compared with those grown in test tubes. By diffusing the reagent into a shorter gel column, the growth process was considerably speeded up, but whether higher speed is necessarily desirable is another matter.

O'Connor *et al.* (1968) developed special procedures for monitoring the decomplexing process. Of the large number of crystals grown, between 5 and 10% were optically clear. These were found to be entirely free from silicon in the bulk, though small amounts remained on the surface even after cleaning. Surface cleaning is best achieved by means of hydrofluoric acid (1 part of 48% solution to 10 parts of water), followed by rinsing in hydrochloric acid and acetone (see also Armington *et al.* 1967a, b; Armington and O'Connor 1967). Storage in vacuum is recommended; in air the crystals may become discolored due to the formation of cupric chloride on the surface.

Cuprous chloride can alternatively be grown by the interdiffusion of cupric chloride and hydroxylamine hydrochloride, as mentioned by Torgesen and Sober (1965), but the method has not yet been optimized to the same extent. Fewer experiments have been made with cuprous bromide, but it is known that this material may be grown by analogous procedures (Armington and O'Connor, 1968a, b).

In a manner similar to that described above, silver iodide, which has always proved difficult to grow, can be complexed with potassium iodide (Fig. 1.3.1(b)). The great solubility of silver iodide in potassium iodide solution and its rapid decrease with increasing potassium iodide dilution make this method particularly suitable. The first successful experiments on growth in solution were reported by Cochrane (1967), and in gels by Halberstadt (1967a, b). For the latter, gels were prepared from 7 ml of sodium silicate solution (244 g $Na_2SiO_3 \cdot 9H_2O + 500$ ml H_2O), 8 ml of 2 M potassium iodide, and 15 ml of 2 M acetic acid. The gelling mixture was allowed to set in a 25×200 mm test tube at 45 °C. A solution was made from 57 g AgI, 260 g KI, and 250 ml H_2O, and added carefully when the gel had set. This solution was allowed to diffuse into the gel for about a week, poured off and replaced by water. The temperature was kept at

45 °C throughout. Pale yellow, hexagonal plates appeared within a few hours and grew to about 10 mm in diameter. One side of the crystals was smooth, the other showed ridges along the diagonals between corners of the hexagon and along lines parallel to the sides of the hexagon, suggesting that growth occurs along only one direction of the c-axis. In the course of electron microprobe analysis, no contaminating silicon from the gel was found (to an accuracy of a few ppm), but potassium was found in quantities up to 6% in certain regions of the crystal, chiefly along the valleys between ridges.

Smaller hexagonal plates (about 3 mm) can be obtained by a slightly different and shorter procedure. Gels are prepared as described above, except that a potassium iodide solution weaker than 2 M, or else just water, is used in making the gelling mixture. The addition of the AgI–KI complexed solution on top of the gel results in crystal growth, typically within a week. Similar experiments at room temperature (about 23 °C) with gels initially free of potassium iodide give somewhat different results, inasmuch as crystals grow in the aqueous layer above the gel as well as in the gel. In Halberstadt's experiments, the crystals in the gel appeared as completely clear small pyramids and prisms (about 0.05 mm). The crystals in the aqueous layer were hexagonal pyramids, 5 mm in diameter and 5 mm high. They had good surfaces but were translucent rather than transparent.

In another variant of the silver iodide growth procedure, hydrogen iodide may be used as the complexing agent in place of potassium iodide, provided oxygen is excluded. For reasons which (along with many others) remain obscure, the presence of hydrogen iodide favors the formation of clear pyramids, as opposed to flat hexagons. The complex may be gradually diluted or else poured off after a time and replaced by water..The former procedure is obviously capable of sustaining growth over a longer period of time. The addition of the supernatant complex may be preceded by an intermediate stage during which a 2 M solution of hydriodic acid is diffused into the gel to lower the local pH, and thereby to raise the solubility of silver iodide. This has a small but significant effect in lowering the number of crystals formed at the immediate gel interface.

The system geometries (Fig. 1.3.5) employed for cuprous chloride can also be used for silver iodide, as can the modified double-diffusion method shown in Fig. 1.3.6. Figure 1.3.7 gives an example of a silver iodide growth system, and Fig. 1.3.8 shows some of the single crystals grown. The best are absolutely clear and of gem-like quality. In refreshing contrast with calcium tartrate, silver iodide is of considerable interest, partly in connection with the mechanism of cloud seeding, partly as an ionic conductor, and partly as a medium for studying the mechanism of the photographic process. It has been grown not only in silica gel (as here), but also in starch (Roy, 1931), and agar (Chatterji and Rastogi, 1951).

Gel-grown crystals of the quality shown in Fig. 1.3.8(b) can be readily cleaved. They have been used, *inter alia,* for tests of the crystallographic polarity of β-AgI. One surface consists of Ag^+ ions, the opposite face of I^- ions. Bhalla *et al.* (1971) have shown that only the silver surface permits etch pits to be formed; the iodine surface remains entirely featureless under potassium cyanide treatment; see Fig. 1.3.9.

Silver bromide and silver chloride have been grown by Blank and co-workers (1967) in similar ways, silver bromide being complexed with hydrogen bromide and silver chloride with ammonium hydroxide; see also Blank and Brenner (1971). Decomplexing (of a lead iodide solution in potassium iodide) is also an alternative method of growing lead iodide. The decomplexing procedure is thus a versatile and valuable addition to the repertoire of gel-growth methods, and the geometry employed permits the removal of waste products in a simple and continuous way. For a general discussion of complexing and decomplexing as suitable procedures for promoting crystal growth in gels, see papers by Nicolau, 1980a, b and by Nicolau and Joly, 1980, which also describe the growth of α-HgI$_2$ crystals by

Fig. 1.3.6. Double-diffusion system with fritted disk. After Nickl and Henisch (1969).

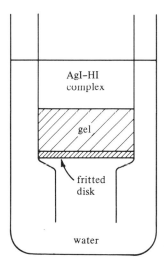

such methods. Nicolau and co-workers actually grew their crystals from solution, but discuss a variety of complexing agents which should function in gels also.

It is also possible to grow crystals by dissolving microcrystalline material in, say, an acid and allowing the solution to diffuse into a gel medium of a pH at which the solubility is much lower. Antimony sulfur-iodide, an interesting ferroelectric material, is an example. It may be formed, for instance, in concentrated hydriodic acid containing antimony sulfide when the pH is raised by dilution or partial neutralization (Dancy, 1969). See also Section 4.4. Another variant involves the diffusion of alcohol into a gel containing the material to be grown. Triglycene sulfate, another ferroelectric crystal,

Fig. 1.3.7. Silver iodide growth system.

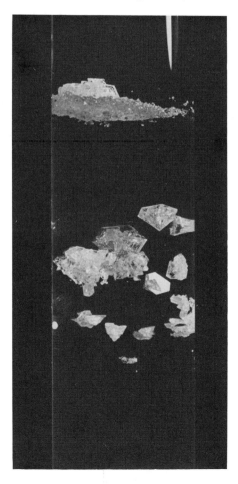

serves as an example for this procedure (Perison, 1968). The material is highly soluble in water, but much less so in alcohol. A gel is made of sodium metasilicate stock solution (see above), and mixed with equal parts of glycene (14 M) and sulfuric acid (2 M). Supernatant ethanol diffuses into this gel after setting, and produces bulky crystals and needles of triglycene sulfate, often with excellent surfaces and optical clarity. The needles may be about 1 mm thick and up to 1 cm long (Fig. 1.3.10). This method has also been used by Glocker and Soest (1969) for the growth of monobasic ammonium phosphate

Fig. 1.3.8. Gel-grown silver iodide crystals. (*a*) Hexagons grown by dissociation of the AgI–KI complex. (*b*) Pyramids grown by dissociation of the AgI–HI complex, 4 mm base diameter. After Halberstadt (1967a, b) and Suri and Henisch (1971).

(*a*)

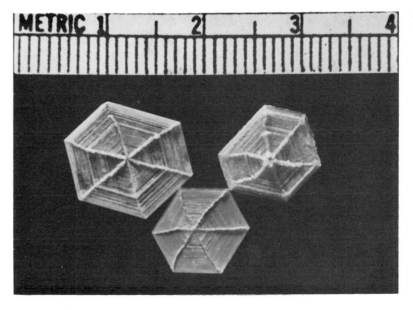

(*b*)

crystals, and by Joshi and Antony (1980) for growing potassium dihydrogen phosphate (KDP). The latter have given a solubility curve which shows that the solubility of KDP in the aqueous mixture drops to about half as the ethyl alcohol concentration increases to 15%, and to one tenth when the alcohol content reaches 40%. This method can therefore be highly effective.

As noted above, the simplest test tube configuration is not always the best. Indeed, despite its disarming simplicity (and in some sense because of it) it is often the worst. On occasions it is therefore useful to employ a double tube, as shown in Fig. 1.3.1(c). The gel may be withdrawn by means of the inner tube, from which it can then be ejected in the form of a continuous column by gentle air pressure. In other circumstances, a double-diffusion system is desirable, either in the shape of a U-tube (Fig. 1.3.1(d)), or as a concentric configuration. Such arrangements are actually necessary for reagents which would form an immediate precipitate when mixed with ungelled sodium metasilicate solution. Because double-diffusion systems involve more variables, they are harder to optimize (but also more rewarding). In principle, not only double but multiple-diffusion systems can be devised, e.g. for doping purposes (see below) or for the production of mixed crystals.

Fig. 1.3.9. Scanning electromicrograph of a cleaved, gel-grown β-AgI crystal. Surface consisting of Ag^+ ions after etching with KCN. The opposite face, consisting of I^- ions, remained entirely featureless. After Bhalla *et al.* (1971).

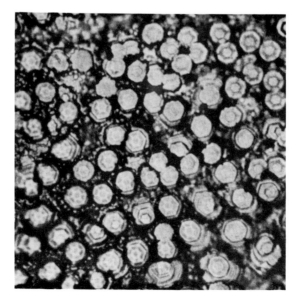

Many of the systems described above lend themselves readily for demonstrations, but large and near-perfect crystals are not generally achieved without the sweat and tears involved in at least some degree of parameter optimization. This is a highly rewarding, generally lengthy, and, at least for the present, largely empirical process, though various guidelines will be found in the chapters that follow. Moreover, it is clear that the crystals in any given system grow competitively, and it is therefore necessary to limit nucleation, an aspect which receives special consideration in Chapter 4. In the absence of nucleation control, there is a greater chance, statistically, of finding a really large crystal in a large growth system than in a small one. Thus, for instance, Armington and co-workers (1967b) were able to grow calcium tartrate crystals of 1 cm maximum size in 50 ml test tubes, and crystals of 2.5 cm length and about 1.2 cm diameter in 500 ml containers.

It has been demonstrated by DeHaas (1963) that crystals can, in principle, be grown by the straightforward diffusion of gaseous reagents

Fig. 1.3.10. Growth of triglycene sulfate. After Perison (1968).

into a solution. Without gel, however, the resulting specimens were smaller than those ordinarily obtained by gel methods, presumably because nucleation could not be suppressed, nor could convection be altogether avoided.

A useful survey of growth procedures has been provided by Arora (1981). These procedures and, indeed, the examples given above, fall into four categories: (*a*) the reaction method, (*b*) the chemical reduction method, (*c*) the decomplexing method, and (*d*) the solubility reduction method. The reaction method often involves an ion exchange between two binary reagents, but it can take other forms as well. Thus, a procedure described by Miller (1937a, b) involves the diffusion of bromine (from an aqueous supernatant solution) into a silica gel charged with potassium iodide, and results in the growth of iodine crystals. Bromine is acting as an oxydizing agent, and in that capacity concentrated (up to 16 N) nitric acid works even better. Miller made similar observations using a variety of other media, including organic gels made by mixing cellulose nitrate solutions in amylacetate with chloroform.

In all cases, irrespective of procedure, the nature of the reaction product should be determined by direct analysis, since double salts and hydrated compounds are often found. For some mysterious reason, unexpected crystals seem to grow more easily than expected ones, and unwanted crystals best of all! Estimated doping levels should be approached with the same degree of caution.

In varying degrees, as mentioned above, all the growth procedures described can be influenced by the application of electric fields, but the resulting field contours are bound to be complex when solute transport by diffusion interacts (via space charge) with solute transport by drift. In systems with low electrolyte content, additional processes may play a role, by analogy with the thermodielectric effect described by Evans (1984). Such processes would depend on the effect to which an electric surface field can influence the probability of 'stiction' of a diffusing (or drifting) crystal component. These matters constitute a fascinating research topic, but have not yet been investigated.

For a better understanding of the processes described above, it is necessary to have some insight into the nature and structure of gel media. These matters are discussed in Chapter 2.

1.4   Purity and doping

There is evidence that the crystals grown in gels can be considerably purer than the reagents used would lead one to expect, a fact which may be connected with the low growth temperature. Armington and co-workers (1967a, 1968) documented the purification effect for Fe, Pb, Cr, Al, and Ca in cuprous chloride, and for Mg, Ni, Na, and Sr in calcium tartrate (1967b). However general purification rules have not yet been established and, perplexing as it may seem, some enrichment effects are also known, e.g. as described by Dennis and Henisch (1967) for iron in calcium tartrate, and by Nickl and Henisch (1969) for various dopants in calcite. All these matters, and specifically the question of how partition coefficients are modified by the presence of the gel, are still in need of further investigation. At the moment there is no safe criterion which enables one to predict whether a given dopant (or contaminant) will be eliminated from or enriched within the growing crystal. The absence, under all normal conditions, of any appreciable silicon contamination may be plausibly ascribed to the general stability of the gel network. However, there are cases (notably calcite) which do not conform to this rule, and these must be dealt with by the hybrid methods described in Section 3.5.

The methods described above can also be used for the preparation of doped crystals, but little systematic work along those lines has been carried out, not even on calcium tartrate, which can in fact be doped with a variety of metals (Ni, Cr, Fe, Co, and Nd). In all these cases the crystals become colored, yellowish green, deep green, yellow, pink, and purple respectively. To some extent the color is a function of the doping level. Thus, 0.065 atomic percent of neodymium has been reported to make the crystals faintly blue, whereas 1.3% makes them purple. Contrary to expectations, the doped crystals were not fluorescent, probably due to the strong absorption of the host lattice.

In the case of calcites, dopants in the form of chlorides or nitrides can be incorporated, before gelling, into the sodium metasilicate solution, or else in the supernatant liquid. Small amounts do not appear to affect the growth habit, but high dopant concentrations prevent the growth of regular rhombohedra. The maximum permissible concentrations for regular growth differ for various metal ions (Nickl and Henisch 1969). For Co, Ni, Cu, Mn, Mg, and Zn, they are of the order of 10 M in the gel, and 5–20 times higher in the supernatant solution. For Cr the maximum is about 0.1 M in the gel. As it happens, all these metals form hydroxides which are insoluble in pure water, but somewhat soluble in the presence of ammonium hydroxide and ammonium salts. For Nd, Ho, Er, and Fe, the maximum permissible concentrations in the gel are of the order of 10 M or, as in the last case, are too small to be readily determined. These metals form hydroxides and carbonates which are even more insoluble (under the growth conditions employed) than those listed above. It is therefore reasonable to assume that they precipitate earlier in microcrystalline form. They would then constitute heterogeneous nuclei (see Section 4.3) on which calcium carbonate growth could take place epitaxially, as suggested by McCauley (1965). Epitaxy would lead to growth forms which have no particular relation to the rhombohedra found at lower dopant concentrations. For the metals which form more soluble hydroxides and carbonates, the same effect would occur, but only at higher concentrations. This interpretation is in harmony with the fact that dopant concentrations which are high enough to destroy regular growth also lead to greatly increased nucleation. Valency considerations must, of course, enter into this picture, but the high permissible concentrations of Cr discount their dominating importance.

There is an enrichment of dopants in the crystals by a factor of about 100, as compared with the average dopant concentration in the gel. This implies that the dopant establishes its own radial concentration pattern round the growing crystals. The volume from which dopant is drawn is thus likely to be many times larger than that occupied by the growth itself. Electron microprobe tests have shown that the dopants are uniformly distributed within the crystals or, at any rate, unclustered, and this is true for calcite crystals clouded by silica network, as well as those free of silica. Of the crystals doped with the above elements, only those containing Mn were found to be photoluminescent (orange) and cathodoluminescent (red).

Systematic experiments on the doping of crystals other than calcite do not appear to have been carried out, nor are many tests on record for the stoichiometric balance of gel-grown crystals. However, Shiojiri and co-workers (1978) did find that some gel-grown, hexagonal rods of thallium iodide were deficient in thallium content. A great deal of work remains to be done in this field.

2

Gel structure and properties

2.1   Gel preparation and properties

Although it is true that good crystals can occasionally be grown in substances that are not normally classified as gels, the general observation is that gels and, in particular, silica gels, are the best and most versatile growth media. Their preparation, structure, and properties therefore deserve attention. At the same time, it is useful to note that no clean-cut demarcation lines between gels, sols, colloidal suspensions, and pastes have ever been established. Standard descriptions of these materials are certainly available but they are not nearly as crisp as one would wish, and many practical substances must be regarded as borderline cases. Thus, for instance, Lloyd (1926) wrote disarmingly that 'a gel is easier to recognize than to define', and even 23 years later the best available characterization referred to a gel as 'a two-component system of a semi-solid nature, rich in liquid' (Alexander and Johnson, 1949). No one is likely to entertain illusions about the rigor of such a definition.

The materials which are ordinarily called gels include not only silica gel (e.g. as usually grown from sodium metasilicate solution), but also agar (a carbohydrate polymer derived from seaweed), gelatin (a substance closely related to proteins), soft soaps (potassium salts of higher fatty acids), a variety of oleates and stearates, polyvinyl alcohol, various hydroxides in water, and even (water-insoluble) tetraethoxysilane in the presence of electrolytes and co-solvents (e.g. methanol) or surface-active agents (Caslavska and Gron, 1984). The materials which are closest to gels in structure are sols; these are also two-component systems, but resemble liquids more than solids. There are also hybrid media which consist of small jelly-like particles separated by relative large tracts of liquid phase. These are sometimes called 'coagels'. In other cases (e.g. hydrated strontium sulfate), the gels appear to consist of crystalline needles in bundles (Hedges, 1931). Resistance to shear can be used as a semiquantitative criterion for the comparison and classification of gel materials. Because crystals can grow in a variety of gels and gel-like media, precise differentiations may not be very important in the present context.

Gels are formed from suspensions or solutions by the establishment of a three-dimensional system of cross-linkages between the molecules of one component. The second component (most commonly water) permeates the network as a continuous phase. A gel can thus be regarded as a loosely interlinked polymer. When the dispersion medium is water, the material should be called a 'hydrogel', to distinguish it from the brittle solids which are often obtained by subsequent drying (e.g. 'silica gel'). In practice, the distinction is not always made; the meaning is usually clear from the context.

The gelling process can be brought about in a number of ways, sometimes by the cooling of a sol, by chemical reaction, or by the addition of precipitating agents or incompatible solvents. Gelatin is a good example of a substance which is readily soluble in hot water and can be gelled by cooling, provided that the concentration exceeds about 10%. In smaller concentrations, the mixture remains a quasi-liquid, the number of cross-linkages being evidently insufficient to establish a recognizable gel. In a similar way, non-aqueous gels can be prepared by cooling sols of aluminium stearate, oleate, or naphthenate in hydrocarbons. On the other hand, Alexander and Johnson (1949) quote some substances (e.g. certain cellulose nitrates in alcohol, and methyl celluloses in water) which obstinately show the opposite behavior: they gel on being warmed!

Gelling by the addition of incompatible solvents or precipitating agents is likewise a simple matter. Solutions of ethyl cellulose, cellulose aceto-stearate, or polystyrene in benzene, for instance, can be gelled by rapid mixing with ether, in which the substances are less soluble. Similarly, a gel of dibenzol-cystine can be prepared by dissolving the material in alcohol and pouring the solution into water (Caslavska and Gron, 1984). Gels of aluminium and ferric hydroxides, vanadium pentoxide, and bentonite can be made from aqueous suspensions by the addition of suitable salts, e.g. $MgCl_2$, $MgSO_4$, or KCL. Although gelatin gels (see above) do not need such additions, their firmness and transparency depend on the pH of the solution before gelling and on the nature of the ions present. A variety of practical gel preparation 'recipes', e.g. for oil, nitro-cellulose, aniline nitrate, and pectin gels, have been given by Bartell (1936). Most gels are mechanically and optically isotropic, except when under strain. However, according to Thiele and Micke (1948), the presence of high ion concentrations can bring about the formation of anisotropic gels (even in the absence of strain) by the alignment of non-spherical sol particles. The same is true for gels prepared from tetraethoxysilane.

Gels can also be formed by the action of two reagents in concentrated solution, e.g. barium sulfate made from barium thiocyanate and manganese sulfate. Indeed, silica gel formation, resulting from the reaction between hydrofluorosilicic acid and ammonium hydroxide, was first described as long ago as 1820 (Hauser, 1955). Along similar lines, detailed procedures for making a whole series of silica–alumina gels of varying pore size have been described by Plank and Drake (1947a, b). In these cases, the aluminium salt is dissolved in some acid before mixing with waterglass or sodium metasilicate. Transparent gels are formed within a few minutes. They have complicated ionic adsorption properties and are ordinarily used as cracking catalysts. When crystals are grown in them, they tend to be contaminated with aluminium (Perison, 1968), which is why they are not ordinarily used for these purposes, despite their otherwise attractive flexibility. Fig. 2.1.1 shows examples of crystals (lead hydroxyphosphate) grown in a non-silica gel, in this case a polyacrylamide hydrogel (4% solution, cross-linked with 0.02% N,N′-methylenediacrylamide); see Březina and Havránková (1980). These crystals are of special interest because they are ferroelectric. Lead hydrogen arsenate has been grown in the same way. In the course of this work, it was also shown that gel growth yielded more perfect (dislocation-free,

Fig. 2.1.1. Crystals of lead hydrogen phosphate grown in polyacrylamide hydrogel. (*a*) Average size: $6 \times 1 \times 4$ mm^3. Growth time: 4 months. (*b*) Average size: $20 \times 1.5 \times 8$ mm^3. Growth time: 6 months. After Březina and Havránková (1980).

(*a*)   (*b*)

inclusion-free) crystals than growth from solution; see also Březina and Horváth (1981) and Section 3.2.

The gelling process itself takes an amount of time which can vary widely, from minutes to many days, depending on the nature of the material, its temperature, and history. For silica gel this has been described and documented by Treadwell and Wieland (1930). During a prolonged period after mixing ('incubation time'), the liquid hydrosol remains outwardly unchanged. This is followed by a comparatively rapid and pronounced increase in viscosity and, in due course, by quasi-solidification. Even before this stage is reached, standard viscosity measurements become meaningless because the material is non-Newtonian. Since gelling is a matter of degree, and is hardly ever quantitatively assessed, quoted gelling times are always very approximate.

The mechanical properties of fully developed gels can vary widely, depending on the density and on the precise conditions during gelling. For instance, silica gels with a molecular silica-to-water ratio of 1:30 or 1:40 can easily be cut with a knife; at 1:20 the medium is rather stiff, and at 1:10, friable (Eitel, 1954). Still denser gels exhibit concoidal fracture surfaces similar to glass. The internal surface area likewise depends on the detailed circumstances during gel preparation (Madelay and Sing, 1962). The available information on silica gels is somewhat 'informal', but the mechanical properties of gelatin gels, relevant as they are to applications in photography, have been investigated a good deal more systematically and have a substantial literature of their own (Hedges, 1931, Saunders, 1955, Saunders and Ward, 1955, Johnson and Metcalf, 1963, and Ward, 1954).

It is often reported that reagents diffuse as rapidly through gels as through water, but it has long since been found that this is true only for electrolytes and very dilute gels. It is certainly not true for large molecules (e.g. organic dyes; see below), nor for colloids. One operative parameter is obviously the size of the diffusing particles, relative to the pore size of the gel. Another is the amount of interaction (if any) between solute and internal gel surfaces. Because convection currents are never completely absent in liquid (gel-free) systems, precise comparisons are difficult. Bechhold and Zeigler (1906) first showed that diffusion coefficients become distinctly smaller as gel densities increase and, indeed, nothing else could have been expected. Stonham and Kragh (1966), who studied the diffusion of KBr through gelatin gels, actually reported a linear diminution of diffusion coefficient with gel concentration. No systematic dependence of diffusion on mechanical gel properties was found. Kurihara and co-workers (1962) concluded from their experiments on the diffusion of the sulfate ion in gelatin gels that the effective diffusion constant is, in this case at any rate, controlled by surface adsorption. Corresponding experiments on silica gels do not appear to have been carried out. There is no evidence that the diffusion constants of small atoms and ions are greatly influenced by the silica gel density, as long as that density is low. This makes it tempting to conclude that it is not greatly influenced by the presence of a dilute gel at all. However, not everything diffuses in a gel; as early as 1862, Graham found that electrolytes diffuse rapidly, but colloids do not. Particle size obviously has something to do with that.

In the course of experiments on crystal growth, the need to determine reagent concentrations in gels arises frequently, and the adsorptive properties of gels make this a difficult problem. A variety of substances adsorb on silica hydrogel with ease. It has been shown (Müller, 1939), for instance, that 1 molecule of iron oxide can be adsorbed on every 5 molecules of silica, and that thorium and yttrium form particularly strong bonds (Sahama and Kanula, 1940). The chemisorption of alkali ions has been studied by Köppen (1938) and cannot fail to be of some importance, considering that silica gels are ordinarily made of sodium metasilicate (see Section 2.2). There is also a selective adsorption of organic dyes; stains produced by fuchsin, methyl violet, and malachite green, for instance, cannot be removed by dialysis, whereas acidic dyes can be leached out. In some cases, there is a change of dye color on adsorption, and this has been ascribed to a polarizing effect of the gel surface. There is room for many theories (not to mention specious mathematics) but even greater is the need for more precise and systematic experimentation. What is known to date is sufficient to support an emphatic caution: the *total* and *free* solute contents of a gel may not be the same. Many attempts to arrive at a chemical composition profile of growth systems are on record, e.g. Ostwald Wo (1925), Fricke and Suwelak (1926), Hedges and Henley (1928), but only by traditional chemical (wet) analysis, which does not easily permit firm distinctions to be made. As a result, we have as yet only a very imperfect understanding of the role played by the internal gel surface, and it is unlikely that this role can always be neglected. These uncertainties have also impeded the practical measurement of diffusion coefficients (see Section 2.3). Early work on the subject was presented in a stylish essay by Adair (1920), who favored the indicator method of measuring the position of a diffusion front on the grounds of its total simplicity: it calls only for a test tube and a ruler. Adair regretted that the underlying theory failed to match this simplicity. In any event, there is another question: what exactly does an indicator indicate? In practice, we do not actually have indicators for every kind of diffusant.

Silica gels prepared in the manner described above are only translucent, as distinct from transparent, and for certain types of experiment this is a handicap; however, there is a viable alternative. Thus, Barber and Simpson (1985) and Barber (1986) have described a method for preparing transparent silica gels, one that involves the use of a cation exchange resin (REXIN 101H). The resin (e.g. 100 g) is first kept in a solution (ca 100 ml) of potassium nitrate or chloride (about 1 M), and stirred for several hours, during which time the potassium replaces other cations in the resin. Resin so treated is then separated, transferred to the sodium silicate stock solution, and left again for several hours. This causes the sodium and potassium ions to be exchanged, and when that process is complete, the resin is filtered out. Gels made with the treated stock solution have a substantially improved transparency for reasons which are not yet understood. Their general performance as growth media remains to be assessed, but there is no reason to believe that the ion exchange will impair their function in any way. In passing, it might be noted that ion exchange procedures have also been described in two very different contexts, for calcium oxalate growth by Cody *et al.* (1982), and for iodine growth by Miller (1937b).

In the discussions which follow, and unless otherwise stated, the term 'gel' is intended to mean silica hydrogel (see Section 2.2), but many other gel-like media, including those mentioned above and some more exotic than others, have in fact been used for experimentation. Among these are not only agar and gelatin, but vanadium pentoxide, cerium hydroxide, zinc arsenate, and manganese arsenate, to mention only some; see, for instance, Roy (1931). However, as already noted, none can be thought of as a true 'general purpose medium', though silica gel comes close; it is highly desirable for its stability, but (unless extensively leached) tends to include unwanted ions by virtue of its preparation. Agar does not have that shortcoming but, as a natural substance, is difficult to purify, and hence is much less well defined and reproducible. Moreover, problems of contamination apart, various organic gels (of which gelatin is one) sometimes interact with electrolyte reagents and change their structure in in response. Typically, they swell; see Hedges (1932).

2.2   Gelling mechanism and structure of silica hydrogels

There was a time, during the earlier years of this century, when the structure of gels was a fighting arena for conflicting theories and models. Lloyd (1926) gives an overview of the controversies, partly in order to reconcile ideas of structure with observations of mechanical properties, which suggest that there is no sharp boundary between gels and viscous liquids. There are, of course, many types of gels, and to some, such structural uncertainties may still apply, but for silica gels, which are here of principal interest, they have been largely resolved. However, even silica gel structures depend significantly on the method of preparation, and thus depend in particular on whether the gels are made by the neutralization of sodium metasilicate or by the hydrolysis of siloxanes. See Lefaucheux *et al.* (1986), and below. When sodium metasilicate goes into solution, it may be considered that monosilicic acid is produced, in accordance with the dynamic equilibrium

$$Na_2SiO_3 + 3H_2O \rightleftharpoons H_4SiO_4 + 2NaOH \qquad (2.2.1)$$

and it is generally accepted that monosilicic acid can polymerize with the liberation of water:

$$
\begin{array}{ccc}
\text{OH} & & \text{OH} \\
| & & | \\
\text{HO}-\text{Si}-\text{OH} & + & \text{HO}-\text{Si}-\text{OH} \rightarrow \\
| & & | \\
\text{OH} & & \text{OH}
\end{array}
$$

$$
\begin{array}{cc}
\text{OH} & \text{OH} \\
| & | \\
\text{HO}-\text{Si}-\text{O}-\text{Si}-\text{OH} + \text{H}_2\text{O} \qquad (2.2.2) \\
| & | \\
\text{OH} & \text{OH}
\end{array}
$$

This can happen again and again, until a three-dimensional network of Si—O links is established, as in silica:

$$
\begin{array}{cc}
\text{OH} & \text{OH} \\
| & | \\
\text{HO}-\text{Si}-\text{O}-\text{Si}-\cdots \\
| & | \\
\text{O} & \text{O} \\
| & | \\
\text{HO}-\text{Si}-\text{O}-\text{Si}-\cdots \\
| & | \\
\text{OH} & \text{OH}
\end{array} \qquad (2.2.3)
$$

though, of course, there is nothing in actual gels that corresponds to the naive rectangularity of this representation. As the polymerization process continues, water accumulates on top of the gel surface, a phenomenon known as syneresis. Much of the water found there is believed to have its origin in the above condensation process, and some may arise from purely mechanical factors connected with a small amount of gel shrinkage.

The polymerization begins immediately, i.e. as soon as the silicic acid monomer is formed. The molecular weight is then low, but Hauser (1955) reported that it increases very rapidly with time as the solution is allowed to stand. This also shows itself in the results of dialysis. When a young gel is dialyzed, a good deal of metasilicate can be washed away, but an old gel no longer permits this, because the molecules are too large.

Fig. 2.2.1(*a*) shows that the time required for gelation is very sensitive to pH. Because gelation is a gradual process, there is no unique definition of gelation time, but almost any definition will serve for comparative tests when linked with a standard procedure. Hurd and Letteron (1932) have described such a method based on mechanical gel properties, and Alexander (1953) describes one based on measurement of the reaction rate with molybdic acid. The results in Fig. 2.2.1(*a*) agree with those given by other workers (e.g. Hurd *et al.* 1934), and suggest convincingly that the reaction is ionic in character (contrary to the impression conveyed

Fig. 2.2.1. Gelling process and acidity; sodium metasilicate. (*a*) Effect of pH on gelling time. (*b*) Changes in pH of an originally neutral gel during syneresis. After Plank and Drake (1947a).

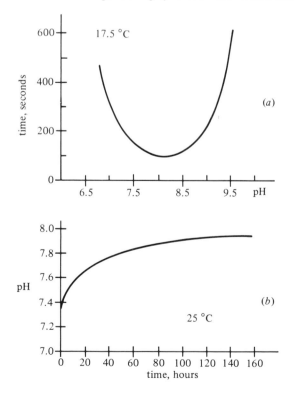

by the simplified representation used above). Some (but, alas, not all) of the electrochemical subtleties have been discussed in a semipopular book by Alexander, *Silica and Me* (1967), so instructively and so charmingly as to make the playful lapse from grammatical grace almost forgivable.

The gelling process is actually more complicated than the above equations suggest, and much is still unclear. However, it is known that, *inter alia* two other ions are involved, $H_3SiO_4^-$ and $H_2SiO_4^{2-}$, produced in relative amounts which depend on the hydrogen ion concentration. The latter, associated with high pH values, is, in principle, more reactive, but the higher charge implies a greater degree of mutual repulsion. $H_3SiO_4^-$ is favored by moderately low pH values and is held to be responsible (Plank and Drake, 1947a, b) for the initial formation of long-chain polymerization products. In due course, cross-linkages are formed between the chains, and these contribute to the sharp increase of viscosity that signals the onset of gelation. The first result of such a linking process would be the production of sol particles, and the extent to which such particles then continue to associate to form a macroscopic gel must depend on their surface charge. Very high as well as very low pH values evidently lead to high surface charges (negative and positive respectively) which inhibit gelation.

Quite apart from questions of charge, it is plausible to assume that very long chains cross-link more slowly than short chains, because of their lower mobility. There is thus a complicated interplay of reactions which, in the absence of other reagents, leads to a minimum gelling time at a pH of 8 or so at 17.5 °C (see Fig. 2.2.1(*a*)). A really precise interpretation of this minimum is not yet available. It appears to be strongly temperature dependent; Alexander (1954) reported that at 1.9 °C gelling is almost instantaneous at pH 6. However, the pH of gelling solutions cannot remain constant; Fig. 2.2.1(*b*) shows, for instance, how the pH of an initially neutral gel increases during syneresis, probably as a result of the progressive and stabilizing hydroxyl substitution for oxygen in the polymerized structure (Plank and Drake, 1947a):

$$-\underset{|}{\overset{|}{S}}i-O^- + H_2O \rightarrow -\underset{|}{\overset{|}{S}}i-OH + OH^- \qquad (2.2.4)$$

In the presence of other ions, the pH changes can be very different (Greenberg and Sinclair, 1955). All this means that an initial pH measurement before gelling can be very misleading as a guide to the prevailing acidity of any gel system after solidification. This is specially so for near-neutral gelling mixtures; in very high and very low pH ranges the effect is not likely to be very important. Greenberg and Sinclair also reported on the fact that the gelling rate is rather sensitively temperature dependent. Some of their results are shown schematically in Fig. 2.2.2. The ordinate is the square root of the turbidity, plotted in this form only because the relationships were then found to be almost linear. The underlying cause is not yet clear; see also Audsley and Aveston (1962).

The formation of cross-linkages can be encouraged by the partial substitution of Al for Si, and because of the difference in valency cross-links form easily. Gelling time is reduced and the resulting gels (when subsequently dried) have a higher density and a smaller pore size than those without Al (Plank and Drake, 1947b).

The well-known stability of silicon-oxygen bonds is responsible for the fact that the polymerization processes described here are largely irreversible, once they have progressed beyond a certain limit. Before that limit is reached, a partially polymerized mixture can be depolymerized by dilution. Hurd and Thompson (1941) have reported on dilute silica hydrogels made with waterglass and acetic acid which could be liquefied by shaking, but it is not clear whether the resulting liquid was actually Newtonian or not; it is more likely that only some of the bonds

Fig. 2.2.2. Dependence of the gelling rate (here measured in terms of turbidity) on temperature. Schematically after Greenberg and Sinclair (1955).

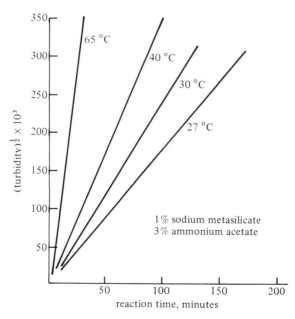

ruptured during shaking. The structure of the silicic acid–water system has been extensively investigated in many other ways, and several comprehensive reviews are available which, in turn, give references to earlier work (e.g. Hauser, 1955, Eitel, 1954, and Iler, 1955). By way of comparison, gelatin gels are believed to be bonded by much weaker forces, possibly of the van der Waals type. Such gels can generally be reliquefied by heating, even without mechanical agitation. In the course of gelling, most systems become somewhat opaque; indeed, light-scattering experiments have been used as an alternative to the methods given above for studying the kinetics of the gelling process (Greenberg and Sinclair, 1955 and Audsley and Aveston, 1962); see also Sections 5.3 and 5.4. For a (highly schematic) attempt at a classification of gel structures see Manegold (1941).

The above representation (eqns (2.2.1)–(2.2.3)) should not be taken to mean that the structural network in silica hydrogel has uniform pores. In fact, it shows only their minimum size, since closed rings of larger size, with and without cross-linkages, can easily be formed. It is therefore reasonable to suppose that there is a distribution of pore sizes within each gel, and that one gel is distinguished from another by the nature of this distribution. In the course of some early experiments on gels containing radium or radiothorium, Blitz (1927) concluded that hydrogels are characterized by two types of pores: 'primary' pores of nearly molecular dimensions, and much coarser 'secondary' pores which behave as more or less normal capillaries. However, whether such a sharp distinction can be made remains open to doubt. In the ordinary way, even the largest pores are too small to be seen under the microscope, though structures are occasionally shown up by ultramicroscopy. Under x-rays, the patterns of silica hydrogel are the same as those observed for silica glass, except for the more intense background which indicates that the gel is less homogeneous (Lloyd, 1926 and Warren, 1940). By observing the progressive sharpening of the interference lines, structural changes which take place during gel ageing can be followed (e.g. see Krejčí and Ott, 1931, Holtzapfel, 1942, and Böhm, 1927).

It is an interesting fact that when a gel contains bubbles, these are usually lenticular in shape, even though they may begin as spheres in the sol. Their orientation is often parallel with one another, which suggests at least some degree of anisotropy and long-range order, unless it can be shown to be due to external factors, e.g. the direction of the maximum temperature gradient during gelling. However, since gels (intended for serious experimentation) are gelled under isothermal conditions, this is probably not the most significant factor. The matter does not seem to have been investigated for silica gels, but Hatschek (1929) has studied it in gelatin by diffusing acetic acid into gels containing sodium carbonate. He found that the bubbles become lenticular quite suddenly during the setting process, an interesting point which deserves to be followed up.

It is in general difficult, though not impossible, to separate the liquid phase by mechanical means, but water extraction by means of drying agents (e.g. concentrated sulfuric acid) has been known for a very long time, e.g. see van Bemmelen (1902). The outcome depends, amongst other things, on the vapor pressure (drying rate) and on the previous gel history (e.g. see Anderson, 1914). Drying has only a small effect on volume, a fact which demonstrates the relative rigidity of the network structure. Dried gels can be more readily subjected to x-ray analysis and other investigational procedures appropriate to solids. As a result, a good deal more is known about them than about the hydrogels from which they are made; and because the fractional volume change on gentle drying is small, measurements on dried gels may yield results which are significant for hydrogels also. On the other hand, the changes which occur during extensive drying (e.g. by heating to 1000 °C) are irreversible and must thus involve the structural network as well as the liquid phase. More studies on reversibility are certainly needed.

With these cautions in mind, it is possible to perform a gravimetric determination of nitrogen adsorption (Plank and Drake 1947a, b and Deryagin *et al.*, 1948) and to calculate the internal surface area from the adsorption isotherms. This, in turn, leads to the establishment of models of the internal structure, and to estimates of the average pore size. Such estimates are necessarily dependent on the assumed pore geometries. They yield effective pore diameters of the order of 50–160 Å for silica gels, and of 28–35 Å for silica–alumina gels of varying density. Pore size is, of course, only one of the descriptive parameters; pore connectivity is another, equally important in the present context and considerably more difficult to define.

As mentioned above, gels can also be made by hydrolyzing siloxanes. Thus, for instance, tetramethylsiloxane (TMS) and water react in accordance with:

$$-\overset{|}{\underset{|}{Si}}-OCH_3 + HOH + CH_3O-\overset{|}{\underset{|}{Si}}-$$

$$\rightarrow -\overset{|}{\underset{|}{Si}}-O-\overset{|}{\underset{|}{Si}}- + 2CH_3OH \qquad (2.2.5)$$

but because TMS is insoluble in water, a good deal of mechanical agitation is needed to bring about this reaction. The great advantage of this procedure is that no additional ions contaminate the system; the methanol produced is harmless. Lefaucheux *et al.* (1986) have described the process and some of the characteristics of the resulting gel media, with TMS contents between 1 and 20% by weight. When these media are dried, they form TMS xerogels, consisting of spherical particles. As for hydrogels, the gelling time depends very much on the presence of foreign ions. Diffusion coefficients have been measured by interferometric holography. For lead nitrate, for instance, the diffusion constant has been shown to diminish rather sharply as the TMS concentration increases from 0 to 15%, and this is, of course, due to the progressive diminution of pore size. TMS xerogels do not (yet) appear to have been studied by scanning electron microscopy.

There is no doubt that the basic gel structure affects the crystal growth characteristics, including nucleation, growth rates, and ultimate crystal size. The mechanisms involved in growth will be more extensively described in Chapter 3, and the effect of gel pore size on nucleation in Section 4.5. For a (highly schematic) attempt at a classification of gel structures see Manegold (1941).

2.3   Gels as diffusion media

A great deal of experimental work has, of course, been done to explore the properties of various gels as diffusion media, but we are concerned here mainly with general principles, and these can, with great advantage, be assessed by computer methods. Analytic solutions of the diffusion relationships are also available, but only for semiinfinite systems. They can sometimes serve as plausible approximations, but give no reliable picture of the (distinctly finite!) systems actually in use.

One looks for solutions of the diffusion equation (derived in Section 5.4):

$$\frac{\partial A}{\partial T} = D_A \frac{\partial^2 A}{\partial X^2} \tag{2.3.1}$$

where $A(X, T)$ denotes the concentration of the diffusing species (A) as a function of position and time. D_A is the diffusion coefficient, which can vary a good deal from gel to gel and from diffusing substance to diffusing substance. Morse and Pierce (1903) estimated the 'ballpark' values to be about 1 cm^2/day (so quoted). Arredondo Reyna (1980) gave values of between 0.2×10^5 cm^2/s and 30×10^5 cm^2/s, in order of magnitude agreement with Morse and Pierce. Typically in the present context, but not necessarily, (A) is an ionic species, one that takes part with (B) in the formation of a compound: (A·B). The exact nature of (A) need not actually concern us, as long as we know that (A) is associated with a simple diffusion coefficient D_A, since many types of system are described by the same equation.

Fortunately, eqn (2.3.1) tends itself particularly well to numerical solution by iterative methods. Thus, it is easy to show (e.g. see Bajpai and co-workers, 1977) that

$$A(X, T+\Delta T) = \tfrac{1}{6}[A(X-\Delta X, T)+4A(X, T)+A(X+\Delta X, T)]$$

(2.3.2)

as long as a certain relationship is maintained between ΔX and ΔT, namely

$$\frac{\Delta T}{(\Delta X)^2} D_A = \tfrac{1}{6}$$

(2.3.3)

This is a restrictive condition, but not one to create any particular impediment in the present context. In the iterative computation, each interval between successive calls upon eqn (2.3.2) represents a ΔT, and the real-world meaning of this ΔT depends in practice on the magnitude of the chosen ΔX, via eqn (2.3.3), which will be recognized as a symbolic representation of the 'drunkard's walk' problem. Eqn (2.3.2) itself means that $A(X, T+\Delta T)$ is simply a weighted average of preceding terms. For the present purposes, it is necessary to fix (or vary externally) only the boundary conditions at $X = 0$ and at $X = L$, where the total length is divided into L segments of length ΔX, so that $X = N\Delta X$. It is convenient here to regard ΔX as unity, which makes N and X numerically equal with $0 < N \le L$. It will be assumed that (A) and (B) have the same diffusion coefficients, which means that they are governed by the same eqn (2.3.3). The algorithm used here (essentially eqn (2.3.2)) allows this condition to be relaxed, but only to a limited extent, with D_B a simple integral multiple of D_A (or vice versa); see Fig. 5.6.10. In the discussions which follow all computational results are presented in dimensionless terms making, in effect, $\Delta X = 1$ and $\Delta T = 1$.

Fig. 2.3.1 gives two widely used system configurations, one with two reagent reservoirs, and one with a single reservoir. The full lines in Fig. 2.3.2 show computed concentration contours for a *semiinfinite* system, for which the diffusion equation can, in fact, be solved by analytic means, the solution being

$$A(X, T) = A_R\{1 - erf[X/2(DT)^{1/2}]\}$$

(2.3.4)

where A_R is the reservoir concentration of (A). In that sense, the full lines serve only to confirm the accuracy of the numerical procedures, since the numerical results approach those calculated by means of eqn (2.3.4) as $L\Delta X$ increases. For want of better solutions, such contours have in the past been used to calculate the positions of precipitates in gels, e.g. see García-Ruiz and Miguez (1982), and Kirov (1978).

It has since been shown (Henisch and García-Ruiz, 1986a) that the differences introduced by taking finite boundaries into account are very substantial, though not, of course, during the initial stages of the process. Fig. 2.3.2(a) illustrates this point. For comparison, the broken line represents the numerical calculation for $T = 100$. As T increases, that line approximates more and more to the chain line, in accordance with obvious expectations: once stability has been reached, with $A(0) = A_R = 100 = $ constant and $A(L) = 0 = $ constant, the concentration must vary linearly between $X = 0$ and $X = L$. Corresponding relationships would apply to any (B) diffusants.

It is sometimes (though not very often) desirable to allow the same reagent to diffuse into a gel from two sides. For such a case, with unequal and constant reservoir concentrations $A_R(0)$ and $A_R(L)$, the situation as a function of time is as shown in Fig. 2.3.2(b). The profile must, of course, become more and more linear as $T \to \infty$.

Fig. 2.3.3 refers to a tube closed at one end (on the right), and its progressive saturation with diffusant (A), coming in from the left. Because there can be no transport through the closed end, the concentration gradient at $X = L$ is always zero. The results are in no way surprising, but are difficult to assess by analytic methods; see Adair (1920) for a solution, brave but unwieldy, in series form. Correspondingly, Fig. 2.3.4 shows how such a gel column, initially charged with solute (A), would

Fig. 2.3.1. Schematic representation of gel-growth systems with uniform gel columns. (a) Two source reservoirs; diffusants (reagents (A) and (B)) reservoir concentrations A_R and B_R. (b) One source reservoir; initial concentration A_G in gel. Length of diffusion system $L\Delta X$. Distance $x = N\Delta X$. It is convenient to take $\Delta X = 1$, making x and N numerically equal, and L the maximum value of N, as above.

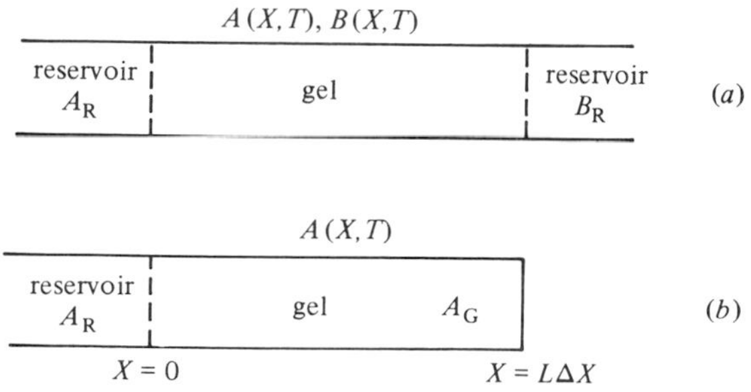

$A(X,T), B(X,T)$

| reservoir A_R | gel | reservoir B_R | (a) |

$A(X,T)$

| reservoir A_R | gel | A_G | (b) |

$X = 0$ $X = L\Delta X$

be depleted (leached), when in prolonged contact with an infinite reservoir of zero solute concentration.

Time-dependent boundary conditions are easily introduced; see also Section 5.6. Thus, in the system to which Fig. 2.3.3 refers, $A_R(0)$ would have to be decremented by an amount proportional to $D_A(dA/dX)_0\Delta T$ for every time interval T, i.e. before every iterative pass of the computa-

Fig. 2.3.2. Computed concentration contours; for a finite (FIN) and a semiinfinite (INF) system (*a*) double diffusion; Comparison of semiinfinite and finite diffusion systems at different times (iterations). Differences show themselves after 30 iterations. $A_R = 100$, $L = 10$ Results for various times T. (*b*) Diffusion of the same component from each side, after different times (iterations). $A(0) = 100$, $A(L) = 50$, $L = 10$. Chain lines: limit at $T = \infty$. After Henisch and García-Ruiz (1986a).

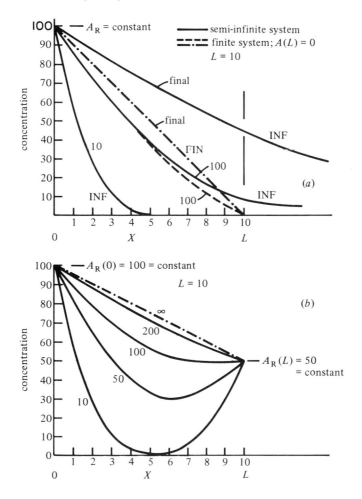

tion. The concentration profiles may be drastically affected thereby, depending on the size of the available reagent reservoir. Some of the consequences of this are discussed in Section 5.6.

One of the quantities of special interest is $A(X, T)B(X, T)$, the actual concentration product, since it is linked to the conditions of precipitation, as discussed in Section 5.5. This product is significant when the diffusion

Fig. 2.3.3. Computed concentration contours for a system closed at one end. $A_R = A(0) = 100$, $L = 10$. Results for three different times T. After Henisch and García-Ruiz (1986a).

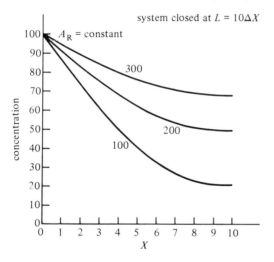

Fig. 2.3.4. Computed concentration contours; leaching of an originally saturated gel by contact with a liquid reservoir of zero solute concentration. $A(0) = 0$, $A_G = 0$, $L = 20$. Results for different times T. After Henisch and García-Ruiz (1986a).

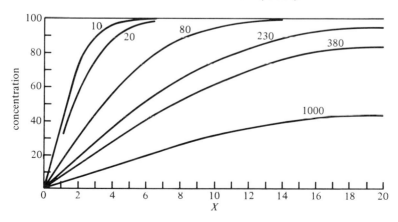

components (A) and (B) form a simple compound (AB); see Section 5.4 for an example in which the product of interest is (A_2B). It can easily be shown on the basis of eqn (2.3.4) that for quasiinfinite systems the AB product is always a maximum at $X = N/2$, independent of A_R and B_R. The numerical solutions in Fig. 2.3.5 show this to be still true for

Fig. 2.3.5. Computed concentration products AB for finite and semiinfinite systems, with equal and unequal boundary concentrations. $L = 10$. Results for different times T. After Henisch and García-Ruiz (1986a).

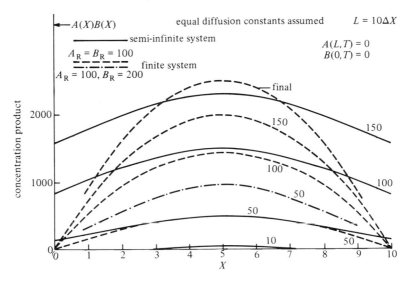

Fig. 2.3.6. Rates of supersaturation increase at the point of first precipitation; comparison between two systems of equal length, one with liquid source reservoirs and one with gel source reservoirs. (Calculation of the first precipitation point is described in Section 5.6.) Double diffusion; system model as shown by the insert. After Henisch and García-Ruiz (1986a).

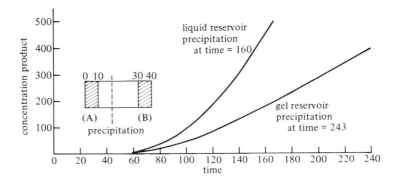

the finite case, as long as (A) and (B) have identical diffusion constants, as assumed here for simplicity. In other respects, the differences between finite and infinite systems will be seen to be pronounced, especially close to $X = 0$ and $X = L$. The matter is important in connection with arguments concerning the position of the first precipitate; see Henisch and García-Ruiz (1986b) and Henisch (1986).

The rate at which the concentration product increases at a place of precipitation can be shown to have an important bearing on the morphology of the precipitation product. Fig. 2.3.6 provides a comparison of two systems, one in which the (A) and (B) reservoirs are themselves gelled, e.g. as they were in some of the experiments by Kai *et al.* (1982), and a more conventional one in which they are liquid, the two systems being equal in all other respects. The results, here only coarsely estimated on the assumption that the precipitation itself withdraws only a negligible amount of solute from the gel, show as expected that the rate of supersaturation increase is much greater when the reagent sources are free of 'internal impedance', i.e. when they are liquids rather than gels.

3

Growth mechanisms

3.1 Diffusion patterns and single crystal growth rates

Gels are obviously permeable, but the fact that convection currents are suppressed, can easily be demonstrated. With an ordinary microscope it is possible to verify that particles have streaming motion in the ungelled solution but none after gelling. With a laser-ultramicroscope arrangement of the kind described by Vand *et al.* (1966), this demonstration can be extended to smaller particles, e.g. down to about 600 Å and even below, depending on the wavelength and intensity of the laser light. Such tests do not rule out the possibility of convection currents on a submicroscopic scale, but it is implausible to believe that these play any major role.

In the absence of convection, the only mechanism available for the supply of solute to the growing crystal or a Liesegang Ring is diffusion. The complete solute diffusion pattern can evidently be very complicated, and attempts to analyze it commit us to several layers of simplification. The choices we make in this must depend, in turn, on the nature of the situation envisaged. One such situation might involve prominent Liesegang Rings, and that will be discussed in Chapter 5. Another might be represented by a small crystal, growing far from anywhere (and, in particular, far from any other crystal) in a large amount of gel. One might then consider that the solute super-saturation ϕ_∞ at large (lateral) distances from the crystal remains unchanged during growth. This would be so, if the total amount of matter in the crystal were small compared with that in the gel, an assumption which is specially appropriate for the initial stages of crystal growth. At the crystal surface, the supersaturation would initially have the same value ϕ_∞ but adjustments to a lower value ϕ_0 would later take place in response to the growth process. It was believed at one time, e.g. see Nernst (1904) and Brummer (1904), that

ϕ_0 corresponds to the saturation concentration (i.e. $\phi_0 = 1$), but it has since become clear that matters are not that simple. For the moment, we shall leave the value of ϕ_0 undefined, and assume only that it would be determined by the dynamics of the growth process itself. We shall return to the matter in Section 3.5, in connection with the problem of cusp formation.

Fig. 3.1.1 shows a schematic representation of concentration contours. Accurate maps would have to be derived by computer and (under realistic and time-dependent conditions) even that is not totally simple; see also Chapter 5. However, the essential features are clear: a radial

Fig. 3.1.1. Diffusion pattern surrounding an isolated crystal (schematic). (*a*) Concentration versus distance X at various radial distances between $R = 0$ and $R = 10.5$ (equivalent to infinity). (*b*) Equal concentration contours.

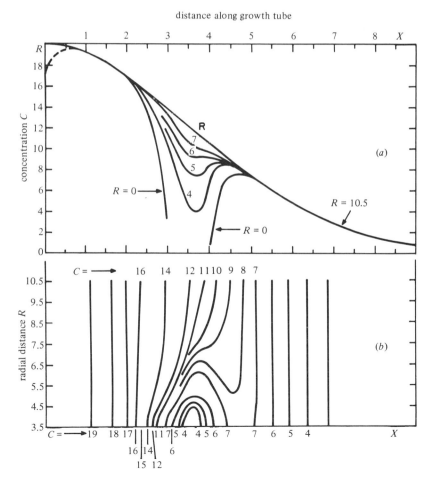

pattern of concentration gradients is superimposed on the main pattern of the gel column.

Immediately below the growing crystal there is a substantial region in which the concentration gradient is small (and, at one point, actually zero); one therefore concludes that the crystal is 'nourished' mainly from above. Accordingly, on those grounds alone, it should grow more rapidly towards the solute source than away from it. Such an anisotropy should be observable with ease, but careful tests do not appear to have been made. An important key issue in this connection is whether the boundary saturation ϕ_0 is indeed a constant, irrespective of the prevailing gradients in the neighborhood. Another is to identify the rate-determining process. This could be the chemical reaction whereby the solute is produced, or else it could be the solute transport between points at which $\phi = \phi_\infty$ and points of $\phi = \phi_0$. Yet another possibility is that the process whereby solute of boundary saturation ϕ_0 finds its way to actual crystal lattice sites might be critically rate-determining. Evidence for the first of these possibilities will be presented below.

If everything were simple, then the concentration contour for large values of radial distance R, far away from the crystal, would be expected to be at least approximately given by eqn (2.3.4), i.e. by

$$C(X_1 T) = A_R \left[1 - \frac{2}{\pi^{1/2}} \int_0^X \exp(-\eta^2) d\eta \right] \tag{3.1.1}$$

with

$$\eta = X/2(DT)^{1/2}$$

D being the diffusion constant (D_A or D_B). This behavior is shown in Fig. 3.1.1(a) by the curve corresponding to large values of R (here 10.5 and effectively equivalent to infinity).

This expectation would follow directly from Fick's 'first' and 'second' diffusion laws, which assume that D is independent of concentration, and that solute flux is conserved as indeed it is, far from the growing crystal; see Shewmon (1963), and also Crank (1956). In this way, by local analyses after a given time of diffusion, D can, in principle, be measured for any gel. Any value so determined would, of course, bear a complex relationship to the gel structure, and would also depend on the nature of the diffusing species. Since we know so little about specific gel structures, D values can, for the time being, serve only for empirical and comparative purposes. It is interesting to note from Fig. 3.1.1(b) that the principal concentration gradient is actually horizontal at one point. Sultan (1952) has observed such relationships experimentally, in solution (sodium chloride growth), using multiple beam interferometry.

Matalon and Packter (1955) have done so by the chemical analysis of gel slices.

In any event, the overall situation is manifestly not simple, because we are not dealing with the monolithic diffusion of a ready-made solute. By the time the diffusion of one reagent, the counter-diffusion of the other, and the reaction between them have been taken into account, the model becomes complex. Indeed, there is no reliable method at the present time that would enable us to calculate the local concentration C as functions of X, R and T. What we do know is that the contour has the general shape shown by the full lines in Fig. 3.1.1(a), except near the liquid interface, where there is a maximum (broken line) due to reverse diffusion, i.e. leaching of the resident reagent into the supernatant liquid. More measurements are needed to ascertain where exactly this maximum is under various conditions, but its visual manifestation is clear enough: it marks, approximately, the first region in which a significant amount of crystal growth takes place. Such maxima arise in many other contexts, and for similar reasons, e.g. in connection with minority carrier injection into relaxation semiconductors, an analogy which comes most pleasingly to the author's mind; see Popescu and Henisch (1975). It shows itself there as a thin region in which the carrier recombination rate is a maximum.

Beyond this, the neat schematic contours of Fig. 3.1.1 notwithstanding, we have little reliable information on concentration gradients surrounding the growing crystal, but new techniques for obtaining such information are being developed. Thus, Bernard and co-workers (1982, 1985) have measured concentration gradient profiles in TMS gels by holographic interferometry, and have confirmed the principal expectations based on Fick's diffusion laws. The pattern of concentration contours must evidently change with time and adjust itself to the needs of the growing crystal surface. When crystal growth domains interact with one another, the diffusion fields become even more complex, and very soon defy all attempts at detailed analysis. In this respect, the interests of the analyst and the crystal grower are the same: both favor growth in isolation.

It is interesting to check the extent to which some of the above notions can be verified by observations on growing crystals as a function of time, even though the verification is bound to be indirect. Frank (1950) has developed equations which give a description of diffusion-controlled growth rates for several different idealized geometries; see also Huber (1959) and references given there. The models are intended to apply specifically to systems in which the growth rate is limited only by volume diffusion, and not in any sense by processes which occur at the crystal surface itself. In solution growth systems stagnancy is hard to achieve, but gels are near-ideal media for experiments under such conditions.

In some (though not all) contexts the essential correctness of the Frank model in terms of concentration contours has already been confirmed by means of multiple-beam interferometry, and the model can therefore be applied with a certain degree of confidence (e.g. see Henisch *et al.*, 1965). The growth rates calculated by Frank involve the 'reduced radius' s, which, for a spherical system, is defined as $r/(DT)^{1/2}$, where r is the crystal 'radius', and T is time. D is the diffusion constant of the reaction product. The theory yields a simple result in the form

$$\phi_\infty - \phi_0 = F(s) \tag{3.1.2.}$$

and Frank did indeed calculate F explicitly for a variety of systems and situations, both two- and three-dimensional. By measuring s and knowing the function F, the value of ϕ_0 at any time could be determined in principle. What is more important in the present case, a constant value of s implies a constant value of the supersaturation difference $\phi_\infty - \phi_0$, and thus of ϕ_0 itself, as long as D remains unchanged. The constancy of s can, of course, be checked by plotting r against $T^{1/2}$ or, since the zero point of the time scale is not known, by plotting r^2 against T (Crank 1956, Sultan 1952).

Some limitations must be borne in mind. One such must undoubtedly arise from the initial transient period during which the steady-state concentration pattern is established; another from exhaustion of the available solute. Yet another could come from the fact that crystals are manifestly not spherical, except in dreams that physicists dream. All these factors are expected to give rise to non-linearity. The duration of the first transition period is not easy to calculate, but the second can be estimated without great difficulty from the original reagent concentrations in the gel and from the distance between the crystal under observation and its nearest neighbor. In addition, one must expect difficulties due to possible disruptions of the gel structure during initial growth, and to changes of pH with time. Nevertheless, and possibly more through good luck than intrinsic merit, tests for the linearity of r^2 versus T have in practice given surprisingly successful results, some in systems which involve only simple crystallization in the presence of the gel, and some which involve a chemical reaction to produce the solute. This strongly suggests that diffusion-transport is often the rate-determining process in crystal growth.

Results of growth rate measurements are shown in Fig. 3.1.2. Each case exhibited significant linear regions, but departures from linearity always set in during the later stages of growth, at times which were plausibly associated with the onset of interaction between the diffusion regions surrounding neighboring crystals. The linear regions themselves may be taken as a rough confirmation of the constant surface-saturation hypothesis. Indeed, considering the tenuous nature of the model, it is remarkable how well the parabolic relationship holds, not only for bulky crystals which might support a quasi-spherical diffusion pattern, but also for needles where such behavior is not really expected. Faust (1968), for instance, has found the same equation to hold for the length of thin dendrites of metallic lead (for growth times between 10 and 200 min). Rate constants may be evaluated from the slopes, and their behavior

Fig. 3.1.2. Crystal size as a function of time; typical relationships. Crystals growing at various distances from the liquid interface. After Henisch *et al.* (1965).

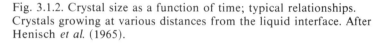

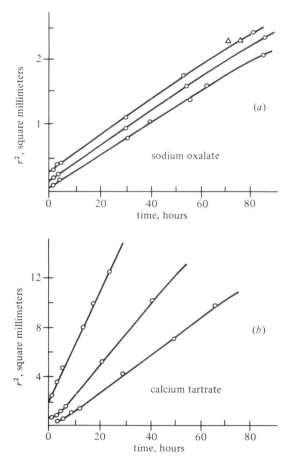

*r*², square millimeters

time, hours

(*a*)

sodium oxalate

*r*², square millimeters

time, hours

(*b*)

calcium tartrate

studied under various growth conditions. The expected departures from linearity during the initial stages of growth are also observed (Fig. 3.1.2(b)) and will be discussed below. Březina *et al.* (1976) have made similar size-versus-time observations in the course of $PbHPO_4$ growth in cross-linked polyacrylamide gels, and Pillai and Ittyachen (1978) in the course of $PbCO_3$ growth.

All this can be done, and the trappings of quantitative research are always enjoyable, but the results should be treated with caution. The notion of ϕ_0, as originally used by Nernst (1904), excludes the possibility of solute concentration gradients parallel to the growth surface. The question is then, do such gradients actually exist and, for gel systems, there is as yet no definitive answer. Observations are on record, for systems involving two-dimensional growth from solution, as a result of extensive experimentation by Berg (1938), and these suggest a more complicated picture. Berg's set-up is shown in Fig. 3.1.3(a), a half-silvered wedge for interferometry, with a layer of liquid surrounding the crystal, sufficiently thin for it to be regarded as two-dimensional and diffusion-controlled. The resulting fringes could be evaluated in terms of super-saturation contours, since the relationship between supersaturation and refractive index was known. The contours, alas, did not in this case confirm the simplistic expectation outlined above. Some typical results

Fig. 3.1.3. Measurement of concentration contours (full lines) and flow-lines (broken lines) round a crystal of sodium chlorate growing in solution, and primarily by diffusion. (a) Wedge system for optical interferometry. (b) Results of observations. Schematically after Berg (1938).

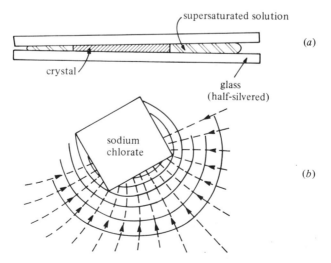

are shown in Fig. 3.1.3(b), which gives flow lines (broken), as well as concentration contours (full). It will be seen that those particular results yield no contour that has the characteristics of the Nernst ϕ_0. We shall return to this matter in Section 3.5, and it will be shown that the notion of ϕ_0 continues to be useful, Berg's results notwithstanding. For the time being, we shall proceed on the assumption that there is, indeed, such a constant.

Meanwhile it is also interesting to look at the more recently observed interference fringes obtained by Lefaucheux and co-workers (1984a and b), as shown in Fig. 3.1.4, which confirm the general pattern of Fig. 3.1.3 to a remarkable degree.

The idea that ϕ_0 may be self-adjusting, adapting itself to the needs of the growth process at every stage, goes a long way towards explaining the high degree of crystalline perfection observed in the course of gel growth; it is one of the distinctive features of the method. It would, of course, be very interesting if the detailed parameters of these systems could be determined by local microanalysis of the medium in the immediate vicinity of the growing crystals. Up to a point (but not elegantly)

Fig. 3.1.4. Observations on the growth of KDP crystals in a TMS gel, by interferometric holography. Photographs (a)-(c) show changes of supersaturation during growth, at half-hourly intervals. Crystal size: 3 mm approx. After Lefaucheux *et al.* (1984a, b).

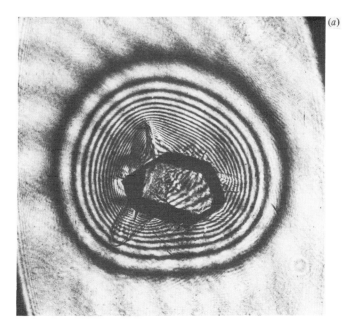

(a)

Fig. 3.1.4 (*cont.*)

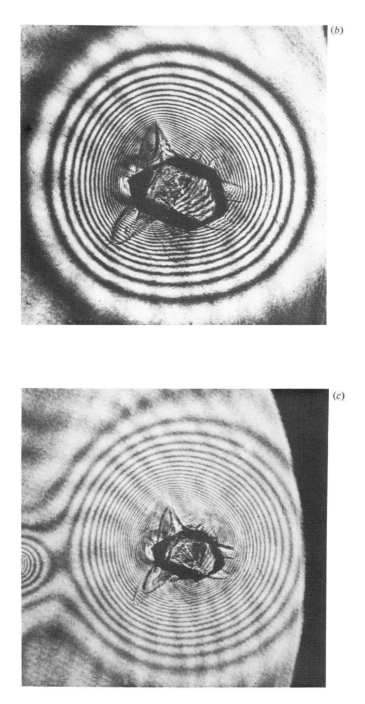

such a thing might conceivably be done by withdrawing samples with a micropipette. In principle, another possibility would be the optical deter mination of concentration gradients in suitably shaped vessels, but the three-dimensionality of the relationships and light scattering by the gel make this difficult.

Growth systems for calcium tartrate crystals permit a further expectation to be correlated with actual findings. Most calcium tartrate crystals exhibit growth veils near their geometrical center. Examples are shown in Fig. 3.1.5. These veils are evidently formed during the initial stages of growth and may well be associated with the non-linearities near the origin of Fig. 3.1.2(b). Such features are not actually peculiar to crystal growth in gels; they are frequently observed in the course of growth from solution. From independent evidence, Egli and Johnson (1963) have ascribed them to 'a growth rate temporarily greater than the crystal can tolerate', and whereas this is less explanatory than one might wish, it is in harmony with the comments above. It is also in agreement with results obtained by Armington and co-workers (1967a, b) in the course of doping experiments: crystals growing rapidly take up more impurity than those growing slowly.

In systems which depend on the diffusion of one reagent through a gel charged with another, the average crystal growth rate is greatest near the top of the diffusion columns, where the concentration gradients are high, and smallest near the bottom, where the gradients are also small.

Fig. 3.1.5. Growth veils near the center of a gel-grown calcium tartrate crystal.

Corresponding to this distribution of growth rates, there is also a distribution in the number of etch pits on any crystal surface, as shown for calcium tartrate in Fig. 3.1.6. If the occurrence of etch pits is accepted as a measure of disorder, in the most general terms, then Fig. 3.1.5 and the arguments above strongly suggest that the growth rate itself determines the number of defects built into the crystal, even in the absence of foreign impurities. Such a process had already been envisaged by Jagodzinski (1963) and Washburn (1958), and is presumably linked with the surface mobility (ordinarily low) of molecules at the growth temperatures in a gel. The average etch pit densities observed on gel-grown calcium tartrate crystals are of the order of 10^3–10^4 pits/cm^2. Occasional specimens have etch pit densities smaller by an order of magnitude, indicating that there is still something to be discovered. The high degree of perfection has also been demonstrated by means of Laue transmission patterns. The general desirability of adopting conditions which ensure slow growth was also recognized by Kirov (1968, 1972).

Once new solute has been brought to the surface by diffusion, growth takes place via screw dislocations (after the initial stages, at any rate;

Fig. 3.1.6. Etch pit count as a function of distance from the liquid interface; calcium tartrate crystals of approximately equal size. After Henisch *et al.* (1965).

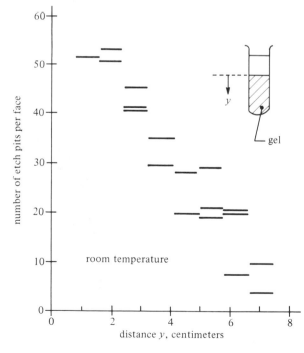

see Section 4.2) or via two-dimensional surface nucleation. One or other of these mechanisms may dominate. Pillai and Ittyachen (1979) found evidence for screw-dislocation-controlled growth on gel-grown calcite crystals, but we do not know how general such findings are. More often, on calcium tartrate, we find structures of the kind shown in Fig. 3.1.7(*a*), with growth layers spreading successively across the face from one side. Their precise origin is not yet clear, but the systematic pattern probably represents growth fronts at right angles to the direction of fastest growth. The steps are in this case prominent enough to be seen under low-power magnification. Fig. 3.1.7(*b*) shows a scanning electron micrograph of a surface on gel-grown silver tartrate, with similar but much finer features. In Fig. 3.1.8, from George (1986), we see gel-grown crystals of metallic silver, with layered cube faces in one case, and clear screw dislocations in another, presumably depending on the exact growth conditions.

The process of getting solute molecules to the active growth front is probably governed by surface diffusion. According to Washburn (1958), surface diffusion coefficients are expected to increase with increasing temperature, leading to greater perfection and thus to fewer etch pits. A small but significant effect of this kind has been observed on calcium tartrate over the practical range of growth temperatures (20–65 °C), though there are other, and possibly more potent, factors which would tend to give the same results (see Section 4.6). It is quite true generally that the crystals which grow in the lower regions of growth systems are larger than those which grow near the gel interface, but proof of greater perfection, as given above for the case of calcium tartrate, has not yet been provided for other crystals. Such information would be valuable, even though, in the last analysis, the features arising from smaller diffusion gradients and those arising from less competitive growth (see below) at greater distances from the liquid–gel interface are not easily distinguished.

Surface nucleation as such has also been extensively studied; the energetic considerations being very similar to those in Chapter 4. The idea is that, in the absence of a screw dislocation, some minimal surface structure must come into being before a crystal can grow (at any rate in a low angle plane). This structure is seen as a monolayer island nucleus, a concept first introduced by Volmer (1939). There is a critical size, and this size diminishes with increasing supersaturation. As yet, calculations of the two-dimensional growth rate do not appear to be in satisfactory agreement with observations, perhaps due to uncertainty about the magnitude of the parameters involved, e.g. see Vasudevan *et al.* (1981). The critical surface nucleus also has an optimum shape which, not surpris-

Fig. 3.1.7. Surfaces of gel-grown crystals. (*a*) Calcium tartrate. Size about 8 mm. (*b*) Silver tartrate under a scanning electron microscope. Carbon replica. After Hanoka (1967).

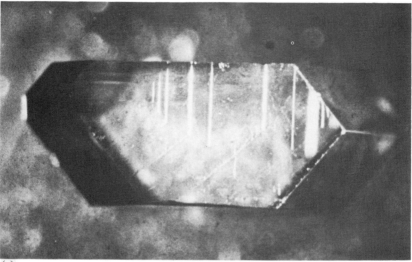

(*a*)

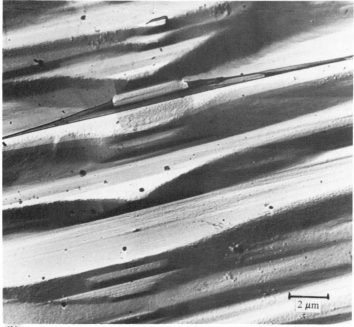

2 μm

(*b*)

Fig. 3.1.8. Surfaces of silver crystals grown by an electrolytically aided gel technique. (*a*) Stepped layer structure. (*b*) Screw dislocations. Contributed by George (1986). See also George and Vaidyan (1981a, b, 1982a, b).

(*a*)

(*b*)

ingly, turns out to be square or circular, no matter what the shape of the subcritical deposit may be. Dhanasekaran and Ramasay (1981a, b, 1982) have calculated that (other things being equal) the free energy required to form such a stable growth center on the surface is actually greater than that necessary for the formation of a classical nucleus; see Section 4.1. For the critical radius, Dhanasekaran and Ramasay (1981b) give

$$r_c = \frac{v_A \sigma}{\Delta \mu_A} = \frac{v_B \sigma}{\Delta \mu_B} \qquad (3.1.3)$$

where v_A and v_B are the partial molar volumes of the (A) and (B) species, σ is the solid–liquid interface tension, and $\Delta \mu_A$ and $\Delta \mu_B$ the changes of chemical potential of the species between inside and outside the nucleus.

3.2. Functions of the gel

The establishment of a stable pattern of concentration gradients as discussed above is regarded as one of the principal functions of the gel. The gel acts, moreover, as a 'three-dimensional crucible' which supports the crystal and, at the same time, yields to its growth without exerting major forces upon it. This (relative) freedom from constraint is believed to be an important factor in the achievement of high structural perfection. Indirect supporting evidence for this view comes from doping experiments. The softness of the gel and, presumably, the uniform nature of the forces which it exerts upon the growing crystal make it possible to overdope specimens until they are metastable as a result of severe internal strain. This can be demonstrated (Dennis, 1967), for instance, by adding a nickel salt to a calcium tartrate growth system. Nickel is accommodated in the crystals, which become green as a result but remain perfectly clear. When the amount of nickel exceeds a certain level, the crystal 'explodes' on contact, audibly if not violently. Nickel contents of the order of 1% produce this effect, which may conceivably have research applications; see Section 3.5.

It is clear that the growing crystal must do some work against the surrounding gel, and attempts have been made to measure 'crystallization pressures'; see below. Inasmuch as the gel structure does, in fact, have to give way to the advancing crystal surface, the pore-size distribution near the growing crystal may be very different from that in the bulk of the gel. Of course, a gel yields easily, but large pressures are sometimes involved when crystals grow in solid porous media, e.g. gypsum crystals in building stones. It is just those pressures that give rise to some of the difficult building conservation problems.

For the measurement of crystallization pressures, Khaimov-Mal'kov
(1958) devised a method of disarming simplicity, as shown by the insert
in Fig. 3.2.1. Seed crystals of aluminium potassium alum were placed on
the top of a gel, with one of the crystal faces in contact with the gel
surface. The gel itself was supersaturated with solute, and as the crystal
grew it was lifted up, eventually to a height of several millimetres. By
placing weights on the crystal, the lift could be suppressed, as shown by
the graph in Fig. 3.2.1(a). In this way, and by monitoring crystal size,
the pressure (i.e. the thrust per unit area) could be simply determined.
This pressure, in turn, was found to be a simple function of the supersaturation (Fig. 3.2.1(b)). For all manner of reasons, some obvious and
some perhaps not, such a measurement cannot be expected to yield
highly accurate results; however, it should certainly be repeated, and

Fig. 3.2.1. Observations of crystallization pressure. Schematically
after Khaimov-Mal'kov (1958). (a) Crystal lift versus pressure.
Insert: nature of the experiment. (b) Crystallization pressure versus
supersaturation. ϕ_∞ = supersaturation far from the growing crystal.
Apparent areas of contact between crystal and gel; 20–90 mm^2 in
various experiments. Crystals of aluminium potassium alum.

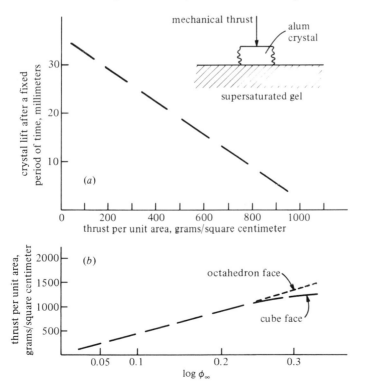

extended to more varied situations than were encompassed by the original work. Meanwhile, the Khaimov-Mal'kov experiment serves as a demonstration of principle, and associates that principle with some numbers that give us an order-of-magnitude insight.

The other important function of the gel is undoubtedly to suppress nucleation, thereby reducing the competitive nature of the growth. Indeed, nucleation control is one of the keys to the ultimate success of the gel method for single crystal growth, and just because nucleation is suppressed, very high degrees of supersaturation can be obtained, without leading to immediate and total precipitation in amorphous form; see Section 3.6. Thus, for instance, Lefaucheux and co-workers (1984a, b) have recorded supersaturations of as much as 26% for KDP in their TMS gels. The problems and opportunities involved are further discussed in Chapter 4. The growth mechanisms operative in the gel after nucleation should not be very different from those in stagnant solutions, though complications may arise in cases in which the gel serves both as a reaction medium and as a diffusion medium.

Very little has been done on the gel-growth of water-soluble crystals, but this subject is of special interest because for such crystals a comparison is possible between the degree of perfection achievable in a gel and that obtained in solution. Lefaucheux *et al.* (1982) have described such studies for TMS gels. In these cases there is actually a choice of methods, inasmuch as crystals can be gel-grown (*a*) by lowering the temperature, or (*b*) by allowing a liquid in which the crystal is less soluble to diffuse into the charged gel medium, or else (*c*) by a combination of both techniques. Fig. 3.2.2 shows a KDP crystal grown by method (*a*). Among the findings are the facts that the growth rates in the gel and in solution are about the same, and that nucleation is suppressed in the gel; see also Section 4.4. In a number of respects (morphology, internal homogeneity) gel-grown crystals were found to be superior to those grown in solution, and their dislocation density was especially low. Similar findings have been reported by Březina and Horváth (1981) for lead hydrogen phosphate (schultenite, $PbHPO_4$) grown in polyacrylamide gel and in aqueous solution. The x-ray topographs of Fig. 3.2.3 show the comparison.

Early workers, e.g. Wi. Ostwald (1897a,b) maintained that the gel behaves as a totally inert medium, but Bradford (1916, 1917) ascribed great importance to the adsorption of reagents on the internal cell walls, without examining that phenomenon in quantitative form, for which, in view of the obvious difficulties, no one will blame him. The literature contains many papers which deal with the question of whether internal surface effects are 'integral' parts of the Liesegang Ring phenomenon (meaning essential to it) or only 'functional' parts (meaning incidental and non-essential). Wo. Ostwald (1926) settled this question once and for all by producing Liesegang Rings in glass capillary tubes, without any gel whatsoever. Of course, the internal surface can play a role, but

Fig. 3.2.2. Crystal of KDP grown in a 7% supersaturated TMS gel by lowering the temperature (35 → 20 °C). Growth duration: 8 days. After Lefaucheux, *et al.* (1982).

it is evidently not an 'integral' one in Wi. Ostwald's sense. Some deft experimentation is still needed to identify and study cases in which that role is prominent. However, because there are many kinds of gels, it is not likely that rules of general validity can be drawn up.

It should be noted that Liesegang Rings grow also in sols (see Sections 4.1 and 5.4) and highly viscous liquids, as Gore (1936,b, 1938a, b) has demonstrated in many different ways. Indeed, they can grow in metals, eg. copper. Gerrard *et al.* (1962) have shown this by allowing Al and As to counter-diffuse. AlAs was found to exist as a metastable phase, deposited in bands. The bands themselves moved with time and ultimately disappeared.

Lastly, Liesegang Rings can grow in air! Spotz and Hirschfelder (1951) observed this, after allowing ammonia and hydrogen chloride to diffuse into an air-filled tube from opposite ends. 'A detailed explanation of this phenomena [*sic*] is difficult', they say. From a knowledge of the diffusion

Fig. 3.2.3. X-ray topographic comparison between lead hydrogen phosphate crystals grown in different media. Mo $K a_1$ radiation. Arrow g indicates the direction on the diffraction vector. Left image pair: polyacrylamide gel-grown; right image pair: aqueous solution grown (a)-images: 100 reflection; (b)-images: 001 reflection. After Březina and Horváth (1981).

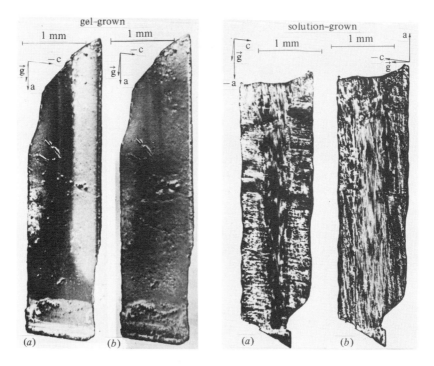

constants, the authors concluded that more than a thousandfold super-saturation was necessary to produce the ammonium chloride precipitate. They also estimated that the critical nucleus (see Section 4.2) contains over a hundred molecules. In passing, Spotz and Hirschfelder also suggest that nucleation can be catalyzed by external irradiation; see Sections 4.5 and 5.2.

Ultimately, the density of the medium, its pore size and its pore connectivity must lead to a diffusion constant or, more accurately, to a set of diffusion constants, one for each diffusing species. Effective constants have been determined straightforwardly (on the basis of eqn (3.1.1)) without counter-diffusion or reaction taking place. Lee and Meeks (1971) did this for a particular agar gel and a large. number of diffusing ions. Their results are reproduced in the Table 3.2.1 below. On the average, the diffusion constants are lower in the gel than in pure water, but not

Table 3.2.1. *Ionic diffusion coefficients*

Ion	$D \times 10^5$ cm^2/s	
	Agar	Water
NO_3^-	0.786	1.92
CrO_4^{2-}	0.75	1.07
Cl^-	0.722	2.03
Pb^{2+}	0.657	0.98
Mg^{2+}	0.638	0.72
MnO_4^-	0.607	
Ca^{2+}	0.588	0.80
Co^{2+}	0.559	
$Fe(CN)_6^{4-}$	0.557	0.74
Zn^{2+}	0.552	0.72
$Fe(CN)_5NO^-$	0.529	
Cd^{2+}	0.525	0.72
Mn^{2+}	0.506	
Cu^{2+}	0.497	0.72
Ni^{2+}	0.493	0.69
$Fe(CN)_6^{3-}$	0.484	0.89
Fe^{2+}	0.479	
Na^+	0.462	1.35
Fe^{3+}	0.434	
K^+	0.264	1.98
MoO_4^{2-}	0.218	

After Lee and Meeks (1971). See also Arrendondo Reyna (1980). Such results are for general guidance only, in view of the difficulty of standardizing gel media.

drastically lower. Unfortunately, comparable data are not readily available for different gel densities, nor for different gels, though they could certainly be established by the same methods. Until such (and other) measurements are made, gel characterization remains one of the weaker points in the interpretational chain.

3.3   Ultimate crystal size; reimplantation

In all simple gel-growth systems (i.e. those without constant concentration reservoirs) the crystals reach a stable, ultimate size, and it is easy to see why this must be so. In fact, three mechanisms are simultaneously at work, and the ultimate crystal size is determined by their interaction. There is firstly the progressive exhaustion of the reagents, which may be simply demonstrated by cutting a cylindrical growth system into disk-like layers and performing the necessary analysis on each layer. Fig. 3.3.1 shows the results of such an analysis for the case of calcium tartrate. In the gel region close to the supernatant liquid, the original tartaric acid content of the gel is completely exhausted, partly through the formation of calcium tartrate crystals and partly through the demonstrable loss of tartaric acid to the solution on top. Secondly as the diffusion process continues and the reaction boundary moves down the column, the diffusion gradients generally diminish (though in Fig. 3.3.1 not very obviously). This tendency is enhanced when the growth tubes are of finite length. Both factors can (and often do) reduce the speed of growth to levels which amount to zero-growth for practical purposes. Meal and Meeks (1968) obtained similar results for the diffusion of $BaCl_2$

Fig. 3.3.1. Calcium tartrate growth; analytic profile of gel column. After Dennis (1967).

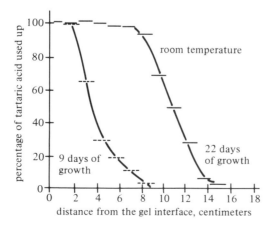

and the counter-diffusion of Na_2SiO_4 in an agar system set up to grow Liesegang Rings of $BaSO_4$.

The third stabilizing factor arises from pH and related considerations. In cases involving the salts of weak acids, of which calcium tartrate is an example, there is a major change towards lower pH values in the course of crystal growth, since the reaction yields rapidly diffusing hydrogen ions. In the increasingly acid environment (down to pH 1), calcium tartrate is increasingly soluble, and a steady state may eventually be reached in which, without any shortage of reagents, growth and solution are in balance. Similar considerations apply to the growth of calcite from (say) $CaCl_2$ and Na_2CO_3. In cases like PbI_2 there is no such acid formation, but PbI_2 is increasingly soluble in the growth medium as the concentration of alkaline KAc increases. This is one of the reaction products, and the fact that it is alkaline favors the formation of PbI (OH) as a secondary consequence; see Section 1.3. This is why thin pale yellow needles of this compound are sometimes found near the bottom of old PbI_2 growth systems.

The researchers' demand for large crystals must come to terms with the question: how large is large *enough*, and the answer depends, of course, on the nature of the experiments contemplated. By means of modern investigational techniques, a great deal can be learned about very small crystals. For most laboratory purposes, crystals of a few millimeters in length are therefore large enough. In a gel system they grow competitively, and as long as they are far apart this competition is ineffective and harmless. However, when nucleation is copious, or else when the crystals have grown enough for their surrounding diffusion fields to interact with one another, this competition becomes a serious impediment. Fig. 3.3.2 gives a (computer-generated) graphic representation of such an interaction; the solute concentration between the two crystals is clearly diminished,

The task of growing much larger crystals therefore resolves itself into the problems of limiting nucleation (see Chapter 4), of ensuring a continued reagent supply, and of removing the waste products. Reservoirs and continuous flow systems can solve the supply problem. Waste product removal is most easily achieved by the decomplexing procedures described in Section 1.3, in which one end of the diffusion column is in permanent contact with distilled water. Corresponding provisions could be made for other growth procedures, though not in the same straightforward way.

When procedures of the kind described above are unavailable or inconvenient, it is still possible to reimplant crystals from an exhausted gel into a new one. In the course of such a transfer, crystals must be handled with great care to minimize surface damage. One method is to place such a crystal onto the surface of a set gel in a tube, and to cover it with more sodium metasilicate solution, which is then allowed to set. Before adding the supernatant reagent, the temperature of the system may be raised temporarily to permit a (possibly damaged) surface layer of the crystal to dissolve. The boundary between new gel and old gel tends to support more nucleation than the gel volume as such. To reduce this problem, the old gel surface should be protected from dust while exposed to air. With these precautions, crystals can be implanted repeatedly, as Dennis and Henisch (1967) have shown, and can be increased in size during each stage. For calcium and copper tartrates this has been done (Bulger, 1969) up to four times, with maximum (total) weight increases by factors up to 19. A statistical spread of these factors is always observed. Similar experiments have been made with calcite, but the method has not otherwise been widely applied, and it remains to be seen how versatile it is.

The criterion for a successful reimplantation is of course, the degree of order at the boundary between old and new growths. This is very

Fig. 3.3.2. Schematic computer simulation of competitive crystal growth. Density of dots (coarsely) represents solute concentration. Vertical overall gradient.

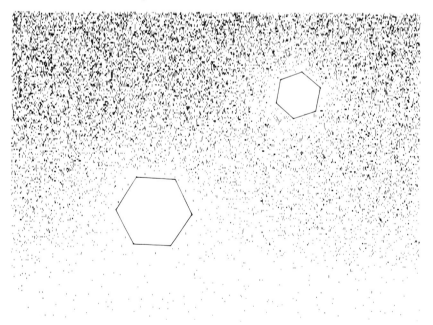

much influenced by the procedures used. Etching before regrowth pro-motes orderly boundaries. It can be performed in two ways, outside or inside the gel (before the addition of new supernatant reagents) and with or without simultaneous heating (see above). For reasons discussed in Section 3.5, etching *in situ* is the more successful process. In calcium tartrate, at any rate, it can lead to growth boundaries which defy visual detection. Imperfect boundaries, when produced, are contaminated by inclusions of sodium and silicon. The latter is, presumably, present in the form of 'trapped' gel. Both contaminants are non-uniformly dis-tributed within the boundary region, which is between 50 and 150 μm thick (Bulger, 1969), and the distributions are not coincident, as may be demonstrated by means of a scanning electron probe. After sufficient *in situ* etching (achieved by simply delaying the addition of supernatant liquid) boundaries on regrown crystals are often free from such contami-nation. To make the *in situ* etching effective, the gel must be sufficiently acidic, e.g. at pH 3. Experiments to determine how gel inclusions depend on gel density and structure have yet to be performed.

It has to be admitted that crystal growth by repeated reimplantation is something less than a crystal grower's dream, but the method is simple and readily available, which is its redeeming feature. See Section 3.6 for growth by hybrid methods.

3.4   Cusp formation

When crystals grow in primary gels, they are often found associ-ated with cusp-like cavities, i.e. regions in which the gel has been split and separated from the growing crystal faces. These cusps may be large and obvious, but often subtle lighting and a microscope are needed to reveal their existence. Fig. 3.4.1 shows what they look like. They arise from the pressure of the advancing growth surface, but whereas the experiment described in Fig. 3.2.1 involves only gel displacement, the present observations concern gel rupture. The resulting cusps were first reported by Khaimov-Mal'kov (1958) who did not, however, draw some of the most important conclusions from the observation. By now such cusps have been seen in a variety of systems, and it is tempting to believe that they are always present, whether directly visible or not.

There is, of course, no doubt that the cusps are filled with solution, but it has long been a matter of puzzlement as to why the crystals should grow just as well where gel is in apparent contact with them as they do where it is not. The contents of the cusps have never been analyzed and it would, indeed, be difficult to do so. For the sake of internal consistency, it is necessary to assume that the solute concentration within them equals that at the growth surface elsewhere on the crystal. This is equivalent to the assumption that solute particles have a high degree of mobility, either on the growing crystal face itself or just above it. The existence of such a high mobility layer has often been proposed in other contexts, e.g. see Volmer (1932), who found this hypothesis essential for the interpretation of his experiments on crystal growth from the melt. Berg (1938), whose results are shown in Fig. 3.1.3, concurred, and for good reason: in the

Fig. 3.4.1. Cusp formation (*a*), (*b*) and (*c*) Lead iodide. (*d*) Calcium tartrate. After Hanoka (1968).

(*a*)   (*b*)

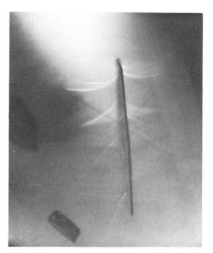

(*c*)   (*d*)

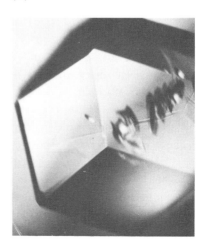

absence of such a layer his crystals could not have maintained planar surfaces during growth.

Faced with the same problem in gels, we are forced to come to the same conclusion, leaving the exact nature of the high mobility layer to be explored. The present case may actually be simpler than first impressions suggest, and some work by Hanoka (1969) clarifies this situation. It will be seen, particularly from Fig. 3.4.1(*c*), that neighboring cusps are, in fact, connected. As a result, the crystal is almost entirely surrounded, not by gel but by solution; and this makes the high mobility and equal concentration hypotheses much more plausible. One might then argue that gel inclusions occur (when they do) only where and when there is actual contact with the gel. The present view, therefore, is that the crystals nucleate in the gel, but grow increasingly from solution, except at a few points of support where, presumably, they rest upon the gel itself. The diffusion process supplies solute to the cusps and, in this way, governs the growth rate. It does not appear likely that convection currents within the small cusp volumes play any appreciable part in the proceedings. Lefaucheux and co-workers (1984a, b) have shown that the presence of cusps makes little difference to the diffusion pattern at substantial distances from the growth surface, but such differences are observed in the immediate vicinity of the crystal–liquid interface; see Fig. 3.1.4. When the growth first begins, and before the cusps have had time to develop, the crystal is, of course, surrounded by gel, and some gel inclusions are then inherently likely, as already noted. Exactly how these inclusions are accommodated is still somewhat of a mystery, but however this may be, as the gel lifts off, the degree of crystal perfection is expected to increase, and this is found to be the case.

The above comments apply to primary growths, in which the development of the crystal and the cusps go hand-in-hand. Reimplantation, followed by immediate growth, disturbs these relationships, since it causes crystals of substantial size to be confronted by an initially cusp-less gel medium. In these cases, the regrowth is governed, at any rate initially, by diffusion through the gel, until new cusps are formed. The initial period after reimplantation is therefore similar to the initial period of primary growth, during which the chances of incorporating gel matter in the growing crystal are greatest. This is in agreement with the reimplantation experiments described above, and may also offer an alternative explanation for the existence of growth veils (of Section 3.1). A crystal implanted and etched *in situ* for a sufficiently long time before growth does not exhibit such disorder, since its initial regrowth is from solution, with only minimum gel contacts involved.

Subsequent growth still differs from ordinary solution growth in some ways which are believed to be important. The surrounding gel permits diffusion which tends to replenish matter taken from the cusps by the growing crystal. It also protects the growth region from the secondary (foreign) nuclei. Moreover, as long as the cusp volume is small enough, the solute concentration is believed to be self-regulating, as discussed in Section 3.1. Accordingly, we are here dealing with a diffusion-controlled process which does not result in fractal growth as, in the absence of cusps, it often does.

3.5   Hybrid procedures; calcite growth

During the initial growth stages, until cusps are formed, the crystal is completely surrounded by gel. However, whereas gel is a good nucleation medium for many types of crystals, it does not follow that it is necessarily as good for subsequent growth. True, the uptake of silicon is, in the ordinary way, so small as to be quite unimportant, but in the case of calcite, for instance, the situation is very different.

A good deal of work has been done on the formation of Liesegang Rings consisting of calcite deposits (e.g. see Gnanam *et al.*, 1980), but little actually on the growth of single crystals. Calcite crystals have well-known applications in optical instrumentation and laser technology; and since the sources of natural specimens appear to be diminishing, a special interest is attached to all methods of growing the material artificially. Previous attempts to grow calcium carbonate by a hydrothermal method have been described by Ikornikova and Butuzov (1956), and experiments on growth from solution by Gruzensky (1967) and Kaspar (1959). Growth in a electrochemical systems has also been investigated (Bárta and Žemlička, 1967). Morse and Donnay (1931) and McCauley (1965) have described growth in gels, the last with emphasis on reaction mechanism and phase aspects. The subject is also of interest in connection with the (painful) formation of calcium carbonate deposits in the human body (Beck and Bender, 1969).

The formation of calcite in sodium metasilicate gels is accomplished by the reaction between carbonates and calcium salts (Hatschek 1911, Fisher and Simons, 1926a, b and Kirov, 1972). Two methods have been developed for this purpose (Nickl and Henisch, 1969). In the first, the gel itself contains the carbonate. An aqueous mixture of sodium metasilicate and a carbonate is prepared, and the pH adjusted to between 7 and 9, usually by means of acetic acid.

After the gel has set, a calcium salt solution is put on top and allowed to diffuse. Attempts to mix the calcium salt with the gel fail because

calcium silicate precipitates at and above a pH of 7. At lower pH values
this precipitation is avoided, but there is a danger of carbon dioxide
production which can destroy the gel. In the second method, the neutral
gel is initially free of calcium and carbonate ions. The reagents diffuse
into it from two sides and form calcite where they meet. This is conveniently done in U-tubes or (better still) in tubes with fritted disk inserts,
as shown in Fig. 3.5.1.

There appears to be no significant difference between the merits of the
two methods; both produce well-shaped calcite rhombohedra of up to
5 mm size within 6–8 weeks. A few spherulites of aragonite and vaterite
also grow. The three structures have been verified by comparison of their
d-values with those compiled by Swanson and Fuyat (1953) and McConnell (1960). Room temperature (25 °C) appears to be optimum for calcite
growth. High temperatures, e.g. 70 °C or so, favor the formation of
aragonite (Kitano 1962, Dekeyser and Degueldre 1950) and cause bubbles
to form which disrupt the gel medium. This should be made from gelling
solutions of analytic grade $Na_2 SiO_3 \cdot 9H_2O$ with concentrations between
0.17 and 0.23 M. As a source of carbonate ions, solutions of Na_2CO_3
(pH 11.6), $(NH4)_2 CO_3$ (pH 9), $NaHCO_3$, (pH 8.6), and NH_4HCO_3, (pH
8.4) may be used; and the calcium can be conveniently derived from
$CaCl_2$ or $CaAc_2$, The combination of $(NH_4)_2CO_3$ and $CaCl_2$ in equal
concentrations (0.16 M) has been found to give the best results. Na_2CO_3

Fig. 3.5.1. Calcite growth in tubes with fritted disks. After Nickl and
Henisch (1969).

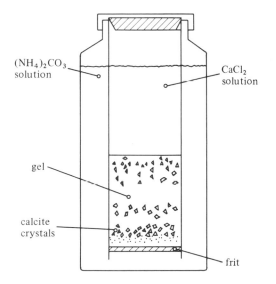

$(NH_4)_2CO_3$
solution

$CaCl_2$
solution

gel

calcite
crystals

frit

is less suitable on account of its high pH, which, upon neutralization, leads to high concentrations of NaAc. This, in turn, affects the quality of the resulting crystals adversely, whereas NH_4Ac appears to have no such effect. See Appendix A1 for a note on the growth of doped $MnCO_3$ crystals.

Crystals grown in the normal way as described are well-formed rhombohedra (Fig. 3.5.2(*a*)) but are almost invariably turbid, obviously due to inclusions. Spectroscopic and microprobe analyses have shown that they contain between 10 and 100 ppm of Mg, which turns out to be the good news. The bad news is that they also contain between 0.45 and 1.7% of SiO_2, when prepared in gels of varying density between 1.02 and 1.03 g/cm^3. Despite this gross contamination, the specimens have well-developed and smooth crystal surfaces. Dissolution of a turbid crystal in acid leaves the internal silica network as a residue which maintains the shape of the original specimen. Fig. 3.5.2(*b*) shows this. The dense inner core is part of the original calcite crystal. The surrounding residue turns out to be silica gel which can be examined by a freeze-drying technique described in Section 4.5, and can be shown to have the same structure as the original gel medium (Fig. 3.5.3). The turbid crystals contain no water. It is clear from the results that the silica network which constitutes the gel is incorporated into the growing crystals more or less intact. In this way, calcite differs greatly from other gel-grown crystals (e.g. calcium tartrate) which are surprisingly free from silica contamination. In these cases the gel is bodily displaced by the advancing growth surface, whereas calcite permeates the silica network, while maintaining a high level of short-range and long-range order. In this respect the gel-grown specimens resemble certain natural calcite structures, e.g. the spikes of sea urchins.

A few crystals are found to grow in fissures, and thus at the boundaries between gel and liquid. These crystals have two regions which can be clearly seen (Fig. 3.5.4); they are turbid where they grow in the gel and clear where they grow in solution. The same observations have been made on crystals of aragonite and vaterite grown by these procedures. This suggests that calcite crystals, and other crystals which may be found to behave in a similar way, should not be grown in gels at all. They should be grown in solution, if possible without losing the general benefits of the gel method. The solution should thus be in a cavity of very small

Fig. 3.5.2. Gel-grown calcite crystals. (*a*) Original form, 3.5 mm maximum size. (*b*) After partial solution in acid. After Nickl and Henisch (1969).

(a)

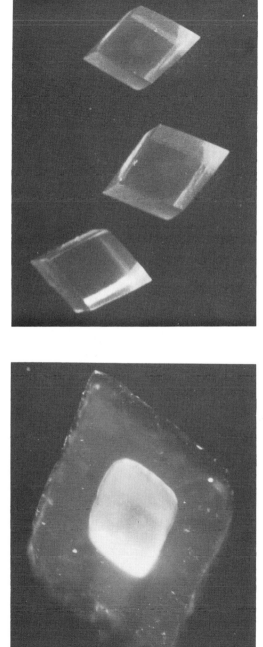
(b)

Fig. 3.5.3. Silica gel structure revealed by a scanning electron microscope after vacuum freeze drying; ×2300. (*a*) Residue after crystal dissolution. (*b*) Normal growth medium. After Nickl and Henisch (1969).

(*a*)

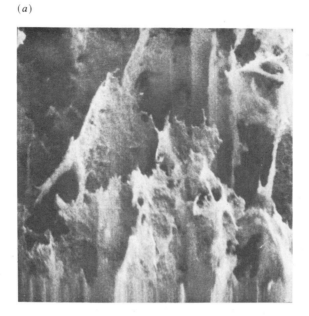

(*b*)

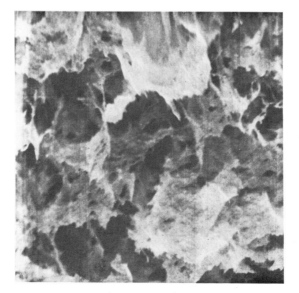

Fig. 3.5.4. Calcite crystals with growth boundaries. (*a*) and (*b*)
Crystals growing partly in solution (*c*) boundaries of gel inclusion.
After Nickl and Henisch (1969).

(*a*)

(*b*)

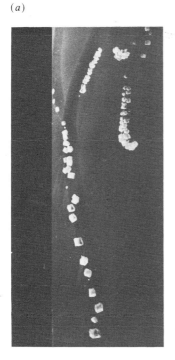

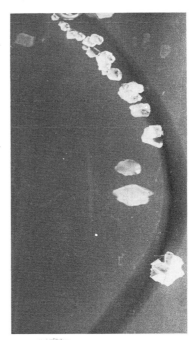

(*c*)

volume, so that the solute concentrations within it may be regulated by the growing crystal. In actual fact, this condition is not easily achieved, and especially not during the intial stages of growth. The solute must be replenished by diffusion in order to achieve the self-adjusting boundary concentration noted in Section 3.1. The diffusion medium must be a gel, so as to prevent secondary (foreign) nuclei from reaching the vicinity of the growing crystal (see Chapter 4). These requirements distinguish the optimum arrangement from the systems, otherwise similar, proposed by Torgesen and Peiser (1968). The solution-filled growth cavity could be seeded with a small calcite crystallite which may originate from a pure gel system but need not do so, especially if clear specimens are available from other sources.

Hybrid procedures of this kind have not yet been widely explored. Fig. 3.5.5 gives two examples which have proved successful, even though they manifestly fail to satisfy the small volume requirement. In the test-tube system (Fig. 3.5.5(a)) sodium metasilicate solution is floated upon a highly concentrated solution of ammonium chloride and allowed to gel before adding the supernatant calcium chloride. Crystals then grow mostly in the solution belt. Unwanted nucleation on the wall of the tube can be diminished by giving it an initial gel coating. The system in Fig. 3.5.5(b) involves diffusion of the reagents through separate gel columns, and the growth medium itself should also be prefiltered in this way. Epitaxial growth on the seed occurs, and the new layers are clear, whether the seed contains a silica network or not. When clear seeds are used, no

Fig. 3.5.5. Systems for calcite growth hybrid gel methods. After Nickl and Henisch (1969).

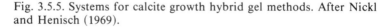
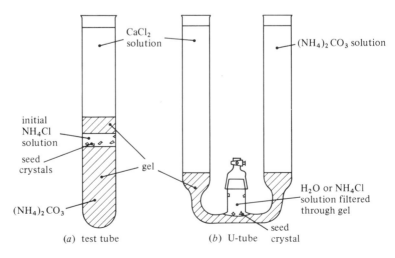

(a) test tube   (b) U-tube   seed crystal

boundary between substrate and new growth can be detected. Weight increases by factors up to 10 have been recorded (Nickl and Henisch 1969). As far as is known, crystals large enough for optical applications have not yet been grown, but there is a real possibility that the gel method might yet come to be used for this purpose. A mathematical analysis of schematic diffusion systems, with particular reference to the local super-saturations produced, has been provided by Lendvay (1965).

Nowhere should the impression be given that hybrid methods are specifically linked with the growth of calcite crystals; they are, indeed, quite generally applicable and have been used, for instance, by Arend and Huber (.1972) for the growth of silver periodate ($Ag_2H_3IO_6$). This is a notoriously difficult material, interesting for its anti-ferroelectric properties and low temperature phase transition; see also Arend and Perison (1971). The Arend and Huber installation is shown in Fig. 3.5.6. Two gel column filters are here contained in U-tubes, one on each side. At the center is a thermally jacketed solution growth system, with its seed crystal. Each U-tube is fed from a reservoir in which the hydrostatic pressure can be (somewhat) adjusted and, in principle, programmed. Some parasitic crystals always nucleate, but far from being harmful, these act as an automatic buffer for the supersaturation. In this particular case, the burettes contained HNO_3 (76.8 g/1) and H_5IO_6 (51.5 g/1)

Fig. 3.5.6. Apparatus for single crystal growth by a hybrid method, using gel column filters. After Arend and Huber (1972).

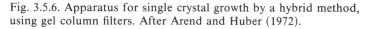

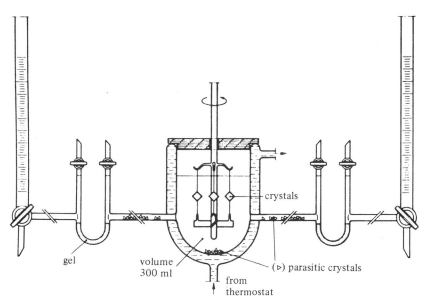

gel   volume   (▷) parasitic crystals
             300 ml
                      from
                      thermostat

respectively, and the periodate was formed by the reaction

$$2AgNO_3 + H_5IO_6 = Ag_2H_3IO_6 + 2HNO_3$$

The growth vessel itself was filled with a saturated solution of $Ag_2H_3IO_6$ in 2% HNO_3. The method could, of course, be more generally applied.

3.6   Lead sulfide; crystal habit

In the laboratory, and particularly in the context of solid state research, one aims at a 'single crystal' of maximum size and exemplary order. In practice, nature does not always oblige, either in the laboratory or in the raw, and a variety of forms may be produced, 'Silica gardens', which owe their existence to the (anisotropic) membrane properties of metal silicates, are an extreme example of the complex, microcrystalline structures which can, in fact, arise; see Coatman *et al.* (1980). In these cases, the preexisting gel plays a role, inasmuch as its constituents take part in the chemical reaction. In contrast, the Liesegang Ring formations described here are 'passive', and involve the gel only as a diffusion medium. Perhaps even more remarkable than silica gardens are the crystalline aggregates produced in biological systems, e.g. see shells ($CaCO_3$), in many of which some degree of long-range order is maintained, despite the basic microcrystallinity; see Amorós and García-Ruiz (1982).

In the context of single crystal growth in gel systems, morphological aspects have been investigated in many contexts, e.g. on metallic silver by George and Vaidyan (1981b, and 1982b) and, perhaps most extensively on PbS, mostly by García-Ruiz and co-workers, e.g. see Aragón *et al.* (1984). The material is of special interest, *inter alia* because it is a semiconductor and, moreover, one in which transistor action has been demonstrated (Banbury *et al.*, 1951). Large crystals are occasionally found in nature, and some of these turn out to be of extraordinary purity. These have, in the past, been the specimens most sought after for electrical measurements, but this may well change.

Room temperature growth procedures have been described by Brenner *et al.* (1966), Blank *et al.* (1967, 1968), and Blank and Brenner (1971), and also by Sangwal and Patel (1974). In all these case, test tubes containing silica gel, acidified by HCl to pH 6, were used, but Aragón *et al.* (1984) found that larger crystals could be grown by using U-tubes, $HClO_4$, and lower values of pH. Hybrid techniques have also been applied; see Section 3.5 and García-Ruiz (1982). Thioacetamine is usually employed as the S^{2-} source. Blank *et al.* (1968) had drawn attention to the low degree of ionization of that compound, a useful characteristic, because it prevents excessive PbS supersaturation from occuring, which might result in abundant nucleation and highly competitive crystal growth. García-Ruiz used $Pb(NO_3)_2$ as the ion donor, either by itself, or else previously complexed with Na-EDTA (where EDTA stands for ethylenediamine tetracetic acid). The reactions were thus:

$$CH_3CSNH_2 + H_2O \rightleftharpoons H_2S + CH_3CONH_2$$

the H_2S being in solution, giving S^{2-}. Also

$$Pb(NO_3)_2 \rightleftharpoons Pb^{2+} + 2NO_3^-$$

or

$$Pb(NO_3)_2 + 2Na - EDTA + H^+ \rightleftharpoons Pb^{2+} + 2NaNO_3 + H\text{-}EDTA$$

and then, of course,

$$Pb^{2+} + S^{2-} \rightleftharpoons PbS$$

Crystals so grown, up to 1.2 mm edge size, were recovered by dissolving the gel with NaOH (1 N). They were identified by x-ray diffraction, and found to be of high purity.

The structure of galena is similar to that of NaCl, i.e. cubic, close-packed, with lead atoms in the octahedral positions. It is, however, believed to be slightly covalent, and the hybrid nature of this binding is expected to have a bearing on the growth morphology, but the main controlling factors are the supersaturation and its rate of change (see below). This is an important respect in which there are great differences between growth in solution and growth in a gel. If high concentrations of $Pb(NO_3)_2$ and thioacetamine were mixed in solution, PbS would quickly precipitate, and the surrounding solute concentration product would very soon adjust itself to $AB = K_s$, an adjustment greatly helped by the prevailing convection currents. Such a medium would be free of macroscopic concentration gradients. As a result, the precipitate would almost immediately be in equilibrium with its surrounding; no further growth would take place. The precipitates could actually be amorphous, since it can be shown that, at very high supersaturations, the size of the critical nucleus (see Section 4.1) becomes smaller than the unit cell of the crystal structure. In a gel, the absence of convection currents prevents the rapid adjustment of concentration. At some stage, and at some place, as the reagents diffuse, the local supersaturation will be high enough for crystal nucleation to take place, but the solute thereby withdrawn would lead only to a very local diminution of the concentration product, and only to a small one, since solute would be continuously replaced by diffusion, in response to macroscopic concentration gradients. As a result, a crystalline deposit, once formed, remains for a while in an environment of high supersaturation. In the course of time, it comes to be exposed to lower supersaturations, which support different growth mechanisms and growth rates. Since PbS is *very* insoluble ($K_s \approx 10^{-28}\,\mathrm{mol^2/l^2}$) its solubility product is quickly reached and exceeded, and even though $AB = K_s'$ is likely to be some orders of magnitude greater, this important point is soon reached; see Section 5.5. In a gel system, the equilibrium situation is actually approached for *two* reasons: (*a*) because the precipitation removes solute from the medium surrounding it; see Section 5.6, and (*b*) because the *overall* concentration contours (which, in turn, influence the local concentration products) approach stability; see Section 2.3.

PbS crystals grow mostly in end forms which are cubes {100} and cubo-octahedra {100}, {111}. While they do so, there is a morphological evolution from skeletal and dendritic crystals to the final forms. According to García-Ruiz (1986), this transformation is dependent on the nature of the medium. Gels acidified with $HClO_4$ support nucleation mostly in the form of dendrites, with more prominent growths at the corners, leading to eight-branched crystals. Later, the open spaces are filled in, leading ultimately to the cube form, with non-flat wavy faces. When the

Fig. 3.6.1. Qualitative model, relating supersaturation and growth habit. After Sunagawa (1981).

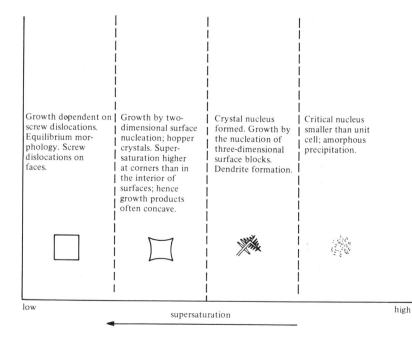

| Growth dependent on screw dislocations. Equilibrium morphology. Screw dislocations on faces. | Growth by two-dimensional surface nucleation; hopper crystals. Supersaturation higher at corners than in the interior of surfaces; hence growth products often concave. | Crystal nucleus formed. Growth by the nucleation of three-dimensional surface blocks. Dendrite formation. | Critical nucleus smaller than unit cell; amorphous precipitation. |

low   supersaturation   high

gel is acidified with HCl, the early growth form is likewise dendritic, but this time along the [100] direction, so that six-branched dendrites are formed. The ultimate 'equilibrium growth' is again in a form close to cube-shaped.

A qualitative model has been established by Sunagawa (1981), which relates growth habit and supersaturation; see Fig. 3.6.1. In general terms, these relationships explain why it is that certain kinds of highly insoluble crystals (of which PbS is one) can be grown in a gel but not in solution. Once a critical nucleus has been created at a very high supersaturation, it grows by three-dimensional surface nucleation, and the result is a dendrite (Fig. 3.6.2(*a*)). As the supersaturation diminishes (for one of

Fig. 3.6.2. Electron photomicrographs of gel-grown PbS; gel acidified with HClO$_4$; progression from (*a*) dendritic crystal grown at very high supersaturation, to (*b*) hopper crystal grown at moderately high supersaturation, to (*c*) polyhedral crystal, grown at low supersaturation. After García-Ruiz (1986).

(*a*)

(*b*)

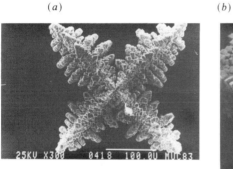

(*c*)

the above reasons, or both) it reaches a critical value at which the growth mechanism changes to two-dimensional nucleation, as proposed by Stranski and Kaischew (1939). This mechanism produces hopper crystals, and the shape of the original structure will therefore change. Fig. 3.6.2(b) shows this for a crystal grown in an $HClO_4$-acidified gel. Finally, the supersaturation will become too low to support surface nucleation, but it will still be sufficient to sustain growth via screw dislocations, in accordance with the Burton *et al* (1951) model. The result will be a polyhedral crystal (Fig. 3.6.2(c)). A corresponding sequence for crystals grown in an HCl-acidified gel is shown by Figs. 3.6.3(a)–(d) and 3.6.4(a) and (b). The principal difference is in the topography of the {111} faces.

Fig. 3.6.3. Electron photomicrographs of gel-grown PbS; gel acidified with HCl; progression $(a) \rightarrow (d)$ for diminishing supersaturation. After García-Ruiz (1986).

(a)

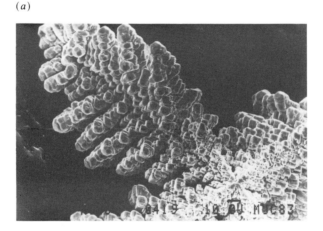

(b)

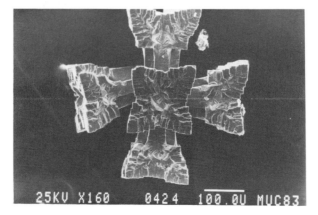

Fig. 3.6.3 *cont.*

(*c*)

(*d*)

The Sunagawa transformation is thus directly observed and indeed not only in PbS; Abdulkhadar and Ittyachen (1977, 1982) report similar observations for lead tartrate and lead halide growths. For lead sulfide García-Ruiz has obtained quite detailed evidence of the fact that three different growth mechanisms are at work. The stage before Fig. 3.6.3(*a*) involves only a six-membered, three-dimensional cross; growth in the [010] directions. Secondary growth families soon appear by block nucleation (Fig. 3.6.3(*a*)). In due course, the spaces between them are

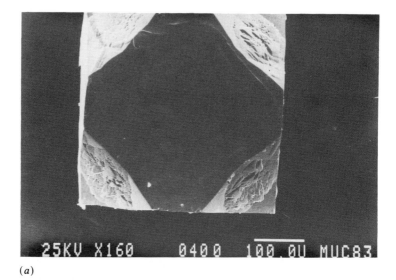

(*a*)

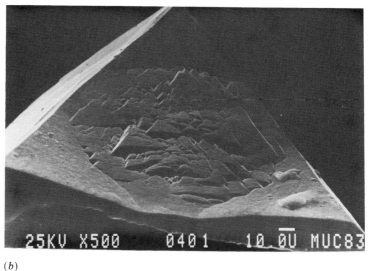

(*b*)

Fig. 3.6.4. Electron photomicrographs of gel-grown PbS; gel acidified with HCl. Final growth stages (*a*). Detail of a {111} face shown in (*b*) After García-Ruiz (1986).

filled in, leading to the shape shown in Fig. 3.6.3(*b*). With more macroscopic filling, that shape changes to Fig. 3.6.3(*c*) and (*d*), the last being closer to the equilibrium form. Fig. 3.6.4 shows the next stage. The {100} faces háve smoothed out; {111} faces are still rough and growing. See García-Ruiz (1986) for crystallographic detail.

4

Nucleation

4.1   General principles

The problem of nucleation is of crucial importance in practical operations, since the crystals which grow in any particular gel system compete with one another for solute. This competition limits their size and perfection, and it is obviously desirable to suppress nucleation until, ideally, only one crystal grows in a predetermined and convenient place. The available techniques do not, as yet, allow us to reach this level of success, though they can sometimes approach it, and sometimes achieve it by happy accident. Growth of a solitary crystal of calcium oxalate, evidently of high perfection, is illustrated in Fig. 4.1.1; see Arora (1981).

Since the application of dislocation theory to these problems (e.g. Frank, 1949, 1950, 1951a, b and Burton *et al.*, 1951) there has been a great increase in our knowledge of the manner in which crystals continue to grow, once growth has started. In comparison, the amount of precise information on the nature of that start is still only small. Always experimentally difficult, the problem is evidently simplest in vapors and melts because only one substance is then involved. It is *a priori* more complex in the case of solutions because of solute–solvent surface interaction and the possibility of nuclei in the course of formation being solvent contaminated (Smakula, 1962).

Crystals growth in gels is evidently a variant of growth in solution, with additional complications arising from the presence of the gel. In this sense, gel systems do not lend themselves well to nucleation studies of the most fundamental kind. Detailed quantitative considerations, though superficially tempting, are therefore not (or, at any rate, not yet) likely to be profitable in the present context. On the other hand, it is now abundantly clear that gels reduce the nucleation probability and, in that sense, offer certain research opportunities which experiments on growth in liquids cannot provide. The nucleation suppressing character (see below) distinguishes gel methods from ordinary diffusion methods sometimes used in crystal growth (e.g. see Lendvay, 1964).

It has long been known that the formation of crystals is sensitively dependent on the presence of impurities. The first systematic and quantitative investigations of this kind are believed to be those of Tammann (1922), mainly on crystallization from the melt. Tammann found that crystallization could be increased by soluble, as well as insoluble (e.g. quartz powder) additives. Since then, two basic nucleation mechanisms have been recognized: *homogeneous nucleation*, which does not fundamentally call for the presence of foreign substances (even though it can be influenced by them) and *heterogeneous nucleation*, which demands a preexisting foreign crystalline substrate (the term with which we often dignify the scientifically tractable part of what is otherwise known as

Fig. 4.1.1. Calcium oxalate crystal. Grown in a system with the configuration shown in Fig. 1.3.4(*b*). After Arora (1981).

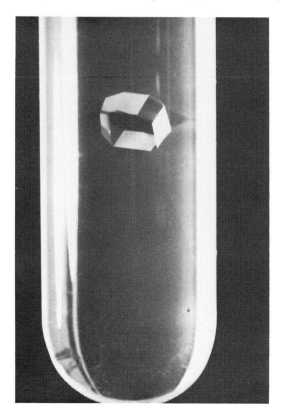

'dust'). On that substrate, new material can be deposited from the vapor, the melt or from solution. One is then dealing with a form of epitaxial crystal growth. It is believed, for instance, that silver iodide serves as an advantageous foreign nucleus during cloud seeding because its lattice constant is close to that of ice (Vonnegut, 1947). Similarly, microcrystals of sodium sulfate are excellent crystallization substrates for sodium carbonate, and phosphates may be used to induce crystallization in solutions of arsenates (Mullin, 1961). The extent (if any) to which non-crystalline particles can serve as foreign nuclei is still unknown. One might guess that they can so serve, but that the new growths, like the substrates themselves, would lack long-range order.

In practice, it is impossible to free systems entirely from foreign particle contamination, and this makes it impossible to guarantee that all the observed nucleation is ever homogeneous. On the other hand, it is feasible (see below) to devise experiments which demonstrate that both types of nucleation can occur and, in particular, that homogeneous nucleation sometimes prevails in gels. Of the two processes, the deposition of solute on a preexisting substrate is energetically 'cheaper'. It therefore occurs at lower supersaturations, and one must expect homogeneous nucleation to be delayed until the heterogeneous nuclei in the system are used up. Because epitaxial growth lends itself more readily to experimentation on a macroscopic scale, its mechanisms are better understood. Even then, there is still uncertainty as to the degree of lattice matching between substrate and deposit that is required for single crystal formation. Moreover, the presence of foreign atoms can affect the mutual binding energy, and can thus influence the heterogeneous nucleation probability. A number of workers have concerned themselves with the manner in which different crystalline substances affect the crystallization of solutes (e.g. see Vonnegut, 1947, 1949 and Volmer and Weber, 1926). The range of possibilities is obviously large, from complete lattice mismatch, on the one hand, to complete matching on the other. An early review of 'catalyzed nucleation', as the process is also called, was given by Turnbull and Vonnegut (1952) and a great deal has been done since then.

All theories of homogeneous nucleation involve the concept of the critical nucleus. It is envisaged that, as a result of a statistical accident, a number of atoms, or molecules can come together and form a rudimentary crystal. Simple energetic considerations show that this crystal is likely to dissolve again (likewise as a result of statistical accident) unless it reaches a certain critical (minimum) size. Beyond that size, the energy relations favor continued growth. This does not necessarily mean that a macroscopic crystal will grow, since this depends, amongst other things, on the availability of sufficient solute. It does mean, however, that the assembly is stable under the prevailing conditions. If the conditions change, e.g. by the appearance of new concentration gradients due to the growth of other crystals in the neighborhood, the growth rate will adjust itself accordingly, and may even become negative. It is a common observation that large crystals grow at the expense of smaller ones, and some of these exploited victims may entirely disappear; see also Section 5.4.

The physical reality of the critical nucleus was first demonstrated by Ostwald (1897a, b); and many attempts have since been made to determine typical sizes. However, because critical nuclei are so small, accurate measurements are not really feasible, and calculated values depend very much on the assumptions made, many of which cannot as yet be independently tested. Calculations (Van Hook, 1963) based on the simplest (though not the most plausible) models lead to critical radii of the order of 10 Å, much too small to be seen even by the electron microscope. The inferences drawn from observations are therefore always indirect.

It is common practice (Mullin, 1961) in the analysis of nucleation phenomena in solutions to rely heavily on analogies with nucleation in the vapor phase. As is well known, small droplets (radius r) of liquid have a higher vapor pressure (p_r) than flat liquid surfaces, a fact expressed by the so-called Gibbs–Thomson formula, which can be written in the form.

$$\log \frac{p_r}{p_\infty} = \frac{2M\Sigma}{R_B t \rho r} \tag{4.1.1}$$

In the equation, M is the molecular weight, Σ the surface energy per unit area (here assumed to be independent of r), R_B the gas constant, t the temperature and ρ the droplet density. All the analogies depend on the notion that p_r/p_∞ can be simply replaced in solution by the concentration ratio C_r/C_∞, where C_∞ is the solute concentration at which 'particles' of infinite radius would nucleate. The nucleation of smaller particles would then demand higher concentrations C_r is accordance with

$$\log \frac{C_r}{C_\infty} = \frac{2M\Sigma}{R_B t \rho r} = \log \frac{\phi_r}{\phi_\infty} \tag{4.1.2}$$

Ideally (but not necessarily in practice), we would have $\phi_\infty = 1$, which led Mullin (1961) to write the equation as:

$$r = \frac{2M\Sigma}{R_B t \rho \log \phi_r} \tag{4.1.3}$$

Both equations express the fact that small solid particles are more soluble than large ones, a fact first analyzed by Ostwald in 1900, and much tested and elaborated ever since (e.g. Jones, 1913, Jones and Partington, 1915a, b, and Dundon and Mack, 1923); see also Section 5.4. In that sense the analogy with the vapor pressures of small droplets is valid. It was, of course, recognized at an early stage that difficulties would arise if the particle radius were allowed to go to zero (not to mention infinity). To avoid this problem, Knapp (1922) postulated that the particles carry a small electric charge which can be shown to reverse the trend otherwise dictated by the supersaturation when r is very small. Further corrections were introduced by Dundon and Mack (1923) by envisaging the dissociation of the solute and various other factors.

Though eqn. (4.1.3) and its elaborations may represent reasonable approximations of the observed facts, they are not in themselves helpful in providing us with deeper insights. It is therefore necessary to limit the argument to more general considerations. The classical argument is that the establishment of a homogeneous spherical (ah, well!) nucleus of radius r *releases* an amount of energy equal to $4\pi r^3 \rho H/3$, where H is the heat of transition involved, in this case the heat of solution. On the other hand, the establishment of a surface *demands* energy to the extent of $4\pi r^2 \Sigma$. Other geometries have been analyzed, but as long as there is no definite knowledge of the manner in which Σ depends on the nature of the crystal surface, such geometrical refinements are unprofitable. The total energy change is evidently a function of r of the form shown in fig. 4.1.2, in which the position of the maximum defines the critical radius r_c, such that

$$r_c = 2\Sigma/\rho H \tag{4.1.4}$$

The critical energy which has to be available for the creation of this nucleus becomes

$$E_c = \frac{16\pi\Sigma^3}{3\rho^2 H^2} \tag{4.1.5}$$

and the nucleation probability itself should be proportional to $\exp(-E_c/Kt)$. Although the average value of E_c must be taken as constant at a given temperature and pressure, local and temporary fluctuations from this average are believed to be possible. At points characterized by temporary minima nuclei are most likely to form.

This is the classical model, and though its quantitative aspects must not be taken too seriously, it remains useful as a basis of qualitative discussions. It is easy to see, for instance, that there should be an optimum temperature for nucleation. As the temperature increases, the probability of E_c being available increases, but the degree of supersaturation diminishes in accordance with eqn (4.1.2). The resulting nucleation maximum was first confirmed by Tammann (1922), not indeed for gel systems, but for supercooled melts. Köppen (1936), many years later, provided similar evidence for KCl solutions. At all temperatures, some of the nucleation is delayed in accordance with stochastic factors, and additional delays may be incurred in viscous media or in systems in which there are other transport problems, e.g. in gels. Under dynamic consitions, this can lead to a substantial build-up of supersaturation.

Of course, models which postulate spherical nuclei cannot by themselves account for the existence of crystals with different growth habits. In some cases, the factors which favor a particular habit have been identified. Thus, when a great deal of heat of crystallization has to be dissipated, dendrite and needle growth are obviously advantageous, because of the large surface-to-volume ratio. In most other cases, the origin of habit-determining factors is still obscure, but there are also systems (notably PbS, mentioned above) that are beginning to be quite well understood; see Section 3.6, and also Sears (1961), who presents a

Fig. 4.1.2. Role of volume and surface energies in the determination of the critical radius r_c. Classical picture; perfect spherical nucleus assumed.

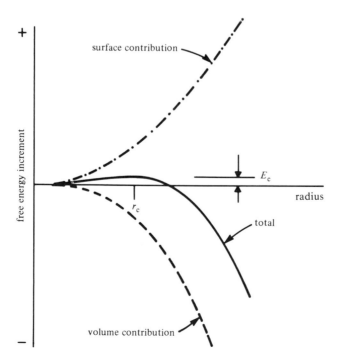

discussion of 'non-classical' nucleation and the origin of spherulites. Many empirical observations have been reviewed by Buckley (1952). Although the crystal habit is evidently determined during the early stages of growth, there is by now ample evidence that the 'habit code' is not irrevocably governed by the structure of the nucleus; it can be influenced by later circumstances.

The effect of macroscopic supersaturation as such can be seen more directly by expressing the above relationships in a different form. Thus, the energy required to establish the nucleus can alternatively be written as

$$E_c = 4\pi r_c^2 \Sigma - \frac{4}{3}\pi r_c^3 \tag{4.1.6}$$

the last term being the work done in creating the volume of the nucleus against a 'pressure' difference Δp, which is known to be $2\Sigma/r_c$. Upon substitution, eqn (4.1.6) becomes simply

$$E_c = \frac{4}{3}\pi r_c^2 \Sigma \tag{4.1.7}$$

The critical radius r_c can then be obtained from eqn (4.1.3) by putting $\phi_r = \phi_{rc}$, this being the supersaturation that produces a stable ('critical') nucleus of radius r_c. Upon further substitution, this yields

$$E_c = \frac{16\pi\Sigma^3 M^2}{3R_B t\rho \log \phi} \tag{4.1.8}$$

where the rc subscript has been dropped, not being actually useful; see also Mullin (1961) and Van Hook (1963), as well as Becker (1983, 1940) and Smakula (1962). Increasing supersaturation thus increases the nucleation probability by diminishing E_c. At the same time, eqn (4.1.8) suggests the possibility of experiments with solutes of different molecular volume. The equation has been amply tested for condensing vapors, and more sophisticated versions of it have also been developed for crystallization from solution, e.g. by Dunning (1955). The interface energy depends, of course, on the solvent as well as the solute, a fact demonstrated in the course of classic experiments by Amsler and Scherer (1941, 1942). These workers crystallized KCl from aqueous solutions which were progressively modified by the addition of alcohol. A corresponding shift in the nucleation probability was duly observed. The effect of impurities on Σ is very complex, inasmuch as Σ increases in some cases and decreases in others. Moreover, the effect which impurities have on nucleation may be different from the effect they have on growth. Apart from the studies described in the following section, there is no record of previous experiments concerned specifically with nucleation in gels.

The entire discussion so far as been based on the idea that supersaturation causes precipitation directly, and that this is indeed the process we see happening in a gel, but this picture has not been accepted without opposition. Dhar and Chatterji (1922), for instance, have maintained that the insoluble material is first produced as a metastable colloidal dispersion, and not as a supersaturated solution. Precipitation is then supposed to occur as a result of coagulation, brought about by the adsorption of additional material on sol particle surfaces. Various observations favor such a hypothesis, but somehow the tests for its validity in particular cases have never been accutely critical. Hybrid models have also been advanced. Thus, Matalon and Packter (1955) favored the sol model but nevertheless drew support in many important respects from Wagner's diffusion analysis. Such a thing is not altogether surprising, since even a sol particle has to be formed somehow, and the above considerations ought to be relevant to that formation. Occasional sol formation is probably part of the general picture, but the widespread and varied nature of crystallization phenomena in gels offers no grounds for believing that sols are *inevitable* precursors of crystal formation.

Because sols may nevertheless play a role, particularly in the development of some Liesegang Ring systems, an outline of these theories is given in Section 5.4. Whether sols play any role in the sparsely nucleating systems discussed above remains uncertain; the salient features of such systems can be explained without recourse to that hypothesis. On the other hand, there are also cases on record (involving barium and lead tartrates) in which macroscopic 'single' crystals were formed *after* the development of a profusely populated, opaque, colloidal suspension in an alkaline silica gel; see Abdulkhadar and Ittyachen (1977, 1980). The gradually diminishing pH, and its effect on adsorption, coupled with the mechanism of competitive particle growth (see Section 5.4) were invoked as explanations. Corresponding experiments in agar did not yield the same results, and this could have been due to the different adsorption properties, as the authors suggest, or else to a different cell size (see Fig. 4.4.4).

4.2   Evidence for homogeneous nucleation

In any reasonably well-optimized gel system, only a few crystals grow to macroscopic size, and the identification of their nucleation mechanism is not a simple matter. That heterogeneous nucleation is possible is no longer in doubt. The phenomenon has been demonstrated by the deliberate addition of foreign nuclei (Henisch *et al.*, 1965) and, in the case of certain (but not all) gel-grown lead iodide crystals (see

Section 4.3), by the detection of silver at the growth center (silver being present in the reagents as a contaminant). The problem is to ascertain whether such nucleation as remains under the cleanest conditions practicable is likewise heterogeneous and thus dependent on very small, possibly subanalytical, foreign nuclei; alternatively, whether such nucleation is homogeneous and dependent on the formation of a native nucleus of critical size, as governed by thermodynamic considerations. Because both processes are expected to depend on solute concentration, pH, impurities, and molecular size, incontrovertible proof is not available, but the following observations (Halberstadt and Henisch, 1968) collectively support the belief that homogeneous nucleation can and does occur in gels.

(a) For the growth of calcium tartrate, it is customary to diffuse calcium chloride from a solution into the silica gel charged with tartaric acid, as described in Section 1.3. However, it is possible to presaturate the growth medium with calcium tartrate by adding a small amount of calcium chloride before the gel sets. Such a gel is clear to the naked eye, but when examined under the microscope using crossed Nicols, numerous crystallites of calcium tartrate are clearly visible. As a result of subsequent calcium chloride diffusion from the supernatant liquid, a very large number of small crystals appear in due course, uniformly distributed over the whole volume. If it were true that these are nucleated by foreign particles, the experiment would show that they are uniformly distributed in the gel as, indeed, one would expect. When the gel is not presaturated in this way, the usual highly non-uniform distribution of crystals develops in the course of diffusion, the number per unit tube length diminishing with increasing distance from the gel–solution interface. In many instances, a good crystal eventually forms in some lower region of the tube, showing that there is sufficient solute concentration in that region, even though a centimeter or two higher, the tube might be entirely free of visible signs of nucleation. Since the concentration of calcium tartrate is bound to diminish with increasing distance from the gel interface, and since heterogeneous nucleation should demand lower supersaturations than homogeneous nucleation, these observations cannot be consistently explained on the assumption that they depend on foreign nuclei. Instead, we must assume that homogeneous nuclei of critical size can ordinarily form only in a few locations in the gel, and even there only in the presence of high supersaturations.

A complication arises from the fact that the concentration gradient is not everywhere parallel to the tube axis, but has radial components in the immediate vicinity of the growing crystal (see Fig. 3.3.2). As the diffusion regions begin to interact, the pattern of gradients ceases to be simple. Moreover, as already noted, large crystals tend to grow at the expense of small ones. The resulting pattern of gradients in three dimensions is then the outcome of stochastic accident.

(*b*) When lead iodide crystals are examined during the initial stages of growth in a gel, they are often (if not always) found to be free from optically detectable defects. Such defects arise during later stages of growth, but it is hard to reconcile the geometrical and structural perfection of the smallest and thinnest crystals with any hypothesis of growth on a foreign substance located (as it would have to be) at the center of the lead iodide hexagons. These crystals are very thin but depend, presumably, on the initial formation of a three-dimensional nucleus. Once a rudimentary sheet structure exists, its thickness might be increased somewhat through the agency of two-dimensional surface nuclei as described by Mott (1950), even before screw dislocations are formed, which ultimately account for all major growth in that (c−)direction. Such surface nuclei may be dirt particles coming into contact with previously perfect growth surfaces. It is believed that the particles need not be crystalline to be effective in this sense.

(*c*) It is a common observation that many different crystals can be grown in a gel, some nucleating very readily and some with difficulty. (Calcium tartrate is a good demonstration crystal precisely because it nucleates only sparingly.) The differences are great, and it is hard to see why foreign nuclei should be so discriminating in their influence on nucleation frequency, while yet supporting in some degree the nucleation of a wide variety of substances. To save the heterogeneous nucleation hypothesis, it would be necessary to postulate that different nuclei are effective for different substances, and that these nuclei are themselves encountered with different though reproducible frequencies in different gels. At that stage the hypothesis becomes too complicated and artificial to be convincing.

(*d*) Some gel-growth experiments are remarkably insensitive to the filtering procedures employed. Although at least some of the foreign particles which are capable of acting as nuclei are presumably too small to be removed efficiently by filtering, some ought to be large enough. It is thus difficult to reconcile the observations in a general way with any theory which ascribes a predominating function to such nuclei; see, however, Section 4.3.

(*e*) Experiments by Kratochvil and co-workers (1968) on the gel-growth of metallic gold crystals (from gold chloride, reduced by oxalic acid) have yielded triangular, hexagonal, and needle-shaped platelets, occasionally with hexagonal pyramids on the growth faces. These configurations (Fig. 4.2.1) bear a striking similarity to vacancy clusters observed in quenched gold prepared by other means. The two structures cannot plausibly be related, except through some process of homogeneous nucleation.

4.3   Studies on quasi-homogeneous and heterogeneous nuclei; filtering

Because foreign nuclei can be very small, their unequivocal detection and identification *in situ* is not an easy matter. As far as is known, it has only once been achieved for crystals grown in gels, in connection with an investigation by Hanoka (1967) and Vand and Hanoka (1967) on the origin of polytypism in PbI$_2$; see also Section 5.1. In some of the growth systems, a cluster of small crystals was found to have formed near the bottom of otherwise only sparsely populated tubes. This suggested that the small crystals might have grown on foreign nuclei heavy enough to have settled before the gelling process or during its initial stages. However, to do this the nuclei would have had to be of a certain minimum size. Estimates indicated that some, at least, should have been large enough to be visible under the microscope, and this was confirmed

Fig. 4.2.1. Photomicrographs of gel-grown metallic gold crystallites. After Kratochvil, Sprusil and Heyrovsky (1968).

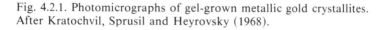

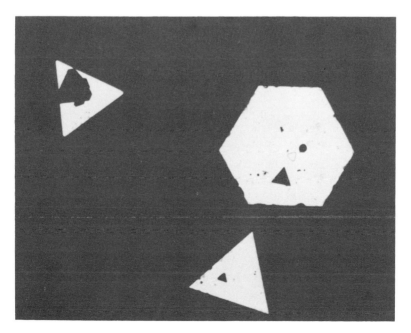

by observations. Some of the PbI_2 platelets had visible spots of sizes up
to 100 μm roughly at their geometrical centers (Fig. 4.3.1). These spots
were first believed to be on the surface, but were, in fact, in the interiors
of the crystals. A slicing technique had to be applied to 'make them
directly accessible for identification by a scanning electron microprobe.
The fact that PbI_2 is cathodoluminescent permitted the centers quickly
to be located under the electron beam, since the emission was noticeably
different at these places. The presence of silver was verified by reference
to the L_α and L_β lines and their comparison with a silver standard. The
silver had its origin in the 'reagent grade' $PbAc_2$ which permeated the
gels, in which it occurred as a notably unlisted but nonetheless prominent
contamination.

The presence of the silver nuclei was found to disturb the crystalline
order in a variety of ways. In some cases it gave rise to six radial lines
of defects emanating from the center and likewise contaminated with
silver. In other cases the silver caused voids, and in some the foreign
nuclei were seen to be located at the center of spirals or at the common

Fig. 4.3.1. PbT_2 crystal containing a foreign nucleus. Field of view =
0.5 mm diameter. After Hanoka (1967).

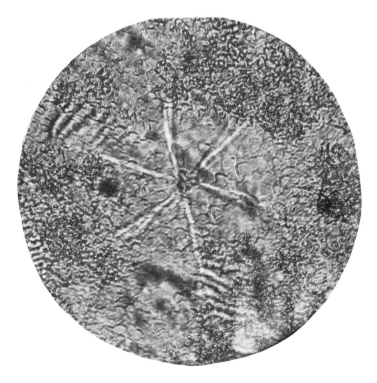

center of a series of hexagons. These findings have an important bearing on the theory of polytypism. They do not prove that all polytypism arises from screw dislocations, but they make it highly plausible that at least some forms of polytypism do, as originally suggested by Frank (1951a, b) and Vand (1951).

In most gel systems there is some growth on the surfaces of the container (usually glass), and experiments on reimplantation (Section 3.2) yield evidence of the fact that boundaries even between two gels solidified at different times are favored as nucleation sites. In a similar way, crystals often grow in contact with macroscopic bubbles in the gel or at internal gel surfaces created by mechanical rupture. All these cases must be regarded as instances of heterogeneous nucleation, but almost nothing is known about the processes involved; see also Section 4.7. To diminish unwanted nucleation arising from gas bubbles, the constituents of the gelling solution should be boiled before mixing.

In an attempt to gain further information about heterogeneous nuclei, it is necessary to explore the effect of foreign additives and the effect of filtering procedures. As far as crystalline additives are concerned, there is some evidence (Henisch *et al.*, 1965), though not of a highly formal kind, that they do increase the number of crystals formed. Controlled experiments are not simple because they depend on the chemical surface history of the dispersed powders. There is also a likelihood that small gas bubbles may adhere to the foreign crystallites, giving rise to highly misleading results. Soluble additives therefore offer more reliable opportunities for observation. Among those tried, e.g. by Perison and co-workers (1968), are various types of sugars. When sucrose, for instance, is incorporated in the gelling mixture, it leads to a sharp increase in nucleation probability. This has been demonstrated for calcium tartrate and silver iodide. Sucrose also affects the gel structure, as may be seen from the greatly increased transparency of the medium. This leaves open the possibility that the effect might be simply 'structural' (see Section 4.5) and not directly due to heterogeneous nucleation. To test this, sucrose was diffused with calcium chloride into an otherwise normal (initially sugar-free) tartaric acid gel. A large increase in nucleation (e.g. by a factor of 10) was observed, and also a change of growth habit, needles being now preferred. In comparative experiments with silver iodide growth systems nucleation was similarly increased. The same holds for the addition of dextrose. When present in the gel from the beginning, dextrose increased silver iodide nucleation by 12% and calcium tartrate nucleation by 100%. When the same dextrose concentrations were employed as supernatant diffusants, the corresponding increases were

412% and 110%. Gel structure changes may be at work, but the results suggest rather strongly that certain organic molecules can themselves act as nucleation centers, an interesting possibility which deserves further exploration.

Experience has shown that the use of ordinary filter papers does not lead to any consistent reduction in the number of crystals grown. Indeed, it has been shown at times to lead to additional nucleation on its own account! The observed inconsistencies are no doubt connected with the fact that cellulose is somewhat soluble in sodium metasilicate solution. Proprietary ('Millipore Solvinert') filters give more consistent results. They too lead to increased nucleation in the case of calcium tartrate, but to significant reduction (factors of about 3) in the number of silver iodide crystals grown. Filters of 1.5 μm mean pore size have proved most effective (Perison, 1968) for the sodium metasilicate solution. Filtering of the potassium iodide solution for incorporation in the gel (using a 0.25 μm mean pore size) produced a further, smaller, nucleation drop. Most significantly, filtering of the supernatant complexed solution had no effect at all. This shows convincingly that diffusion through the gel itself is the most effective fitering process, at any rate as far as heterogeneous nuclei above a certain size are concerned.

Sugar molecules (see above) are evidently smaller than the critical size, and 'Millipore' filters cannot eliminate these, but spectroscopic analysis of the filter plates has shown that at least some foreign iodides are indeed removed from solution during filtering. Substances which manage to pass through can act as nucleation centers when present within the pore structure of the gel. This can be simply demonstrated. When small amounts of silver iodide powder are suspended in the gel during its formation, they lead to an enormous increase in the number of macroscopic crystals eventually developed. Self-epitaxy is, of course, the most likely process, but there is no reason to believe that other iodides (or, indeed, other substances) which are isomorphous with silver iodide, or nearly so, will behave differently. At the same time, the number of visible crystals remains orders of magnitude smaller than the number of powder grains added. The simplest explanation is probably the best: that a great deal of nucleation is suppressed by gel envelopment of foreign substrates, which prevents solute from reaching them in amounts sufficient for macroscopic growth.

The above results suggest that the optimum conditions for nucleation suppression would prevail if all components of a growth system were filtered through gel. This is a simple matter as far as the reagent solutions are concerned, and the success of such procedures is well demonstrated by the hybrid methods described in Section 3.5. However, sodium meta-silicate solution itself does not pass easily through a gel, and a really satisfactory method of filtering it still remains to be found. There remains, of course, the general problem of preparing gels (from commercially available reagents) that are free from impurities in solution, since no filtering process can cope with these. That this is not a minor matter for certain types of work has been noted by many previous researchers, including Riegel and Reinhard (1927), Schleussner (1922), Foster (1918) and, indeed, Liesegang himself (1914).

The present discussion has concerned itself with the role of heterogeneous rather than homogeneous nuclei, but there are situations in which the distinction breaks down. Thus, Hatschek (1914) experimented with lead iodide growth systems in which the gel contained crystalline lead iodide in suspension even before the onset of diffusion. The question, deemed to be highly critical at the time, was whether crystallites would behave like homogeneous nuclei. If they did, and since they were plentiful, how could any macroscopic supersaturation ever arise? In Hatschek's experiments such microseeds were actually found to play very little part, which may have had something to do with the nature of their surfaces. They were clearly behaving as heterogeneous nuclei, inasmuch as they supported nucleation at all. As already mentioned in the discussion of hybrid procedures (Section 3.5), an initial resolution stage is virtually essential if ready-made seeds are to perform well as nuclei. A similar observation is on record for agar gels seeded with microcrystalline lead chromate; see Hatschek (1919).

Without initial resolution, continued crystallization is indeed possible, but it demands solutions of much higher supersaturation. Up to a certain maximum solute content, even highly supersaturated solutions can remain stable in the presence of inactive nuclei, as Morse and Pierce (1903) found. A demonstration of the fact that high supersaturations do, in fact, occur in normal growth systems in the presence of quasi-homogeneous nuclei has been given by Van Hook (1938a, b), who found that it took something like 22 days for a supersaturation of 8 to equilibrate. This confirmed earlier experiments by Bolam (1928, 1930, 1933) and Bolam and Mackenzie (1926), but left unclear whether these long delays were due mainly to the nature of the seed surfaces or to the protective character of Van Hook's gelatin and agar gels. The fact that the results were essentially similar in a variety of gel media suggests the former. Indeed, Ghosh (1930) had already obtained the same results in water, even without a gel.

Van Hook (1983a, b) did, however, find that there was an upper limit to the permissible seed concentration. He argued that the crystal mass m must grow at some rate:

$$dm/dT = n\alpha Sf(\phi) \tag{4.3.1}$$

where n is the concentration of preimplanted seeds, each of surface area S, α is a constant and ϕ the supersaturation. The functional form of f was not actually known, but Van Hook used

$$f = \phi^\beta \tag{4.3.2}$$

to test matters in principle, β being a constant that could be empirically fitted. It was found that $\beta = 2$ worked best, which may or may not have a profound meaning.

The growth front should stabilize where the growth rate equals the rate at which solute arrives by diffusion, i.e. (taking the silver concentration to be equivalent to the solute concentration) at the rate $D_A d^2C/dx^2$; see also Section 5.4. With these provisions the equations can be solved. They yield the result that nSx^2 should be constant for the place along the growth tube at which growth ceases. Van Hook was able to confirm this relationship experimentally for a variety of seeds in greatly varying concentrations. He grew silver chromate, and his seeds were also silver chromate, but because these were in many of his runs prefabricated (externally precipitated), the chances are that they behaved like heterogeneous nuclei.

A complicated and as yet unresolved problem arises from a question first posed by Dhar and Chatterji (1925a, b): to what extent do sol particles (as distinct from single molecules in solution) react to an implanted seed? These experimenters believed that sol formation, followed by coagulation is the normal sequence of events in Liesegang Ring formation, and maintained that sol particles do not react with seeds directly.

4.4   Nucleation control

In the context of purposeful crystal growth, the suppression of nucleation is the principal function of the gel, but it is apparent that the degree of suppression ordinarily obtained is insufficient for many of the crystals one wishes to grow. As it happens, calcium tartrate nucleates only sparingly, and that, rather than any fascinating property of its own, explains why it is so often used for crystal growth demonstrations. It is, in fact, a remarkably unexciting material but precisely because it grows so well in gels it lends itself to a variety of studies on the behavior of gel systems. Many other crystals nucleate more easily and, as a result, grow less well. In such cases, there is every incentive for a search to discover additional means of nucleation control.

In principle, the simplest way of achieving an additional reduction of nucleation would be to accentuate the 'natural' suppressive action of the gel. This, in turn, must have something to do with the pore size distribution and with the extent to which pores are in communication with one another. The simplest model suggests that homogeneous nuclei of critical size cannot form in very small isolated pores because the necessary amount of solute is not available. Critical nuclei may form in larger pores, but they are not expected to grow to larger (macroscopic) size unless there is a sufficient degree of communication with other pores containing solute (or components from which solute is made). Much the same factors would control the effectiveness of heterogeneous nuclei. Some of these might be completely enveloped and 'protected' by gel matter, and would then become inoperative; others might happen to be in suitable locations in which continued growth could be supported by the prevailing diffusion conditions. It could also be that some potential nuclei are 'used up' by the gel structure itself, held in the three-dimensional silica network by chemical bonding forces, and thereby rendered ineffective as crystallization substrates. The results discussed in Section 4.3 show that this happens with sugars, but there is no evidence that insoluble crystalline additives can be built into the gel network in this way.

The present picture ignores any part that may be played by the pore walls, but it suggests that there are favorable and unfavorable pore-size distribution. The comparison between silica and gelatin gels (Section 4.5) bears this out in rough terms, and the model is also in agreement with experiments involving PbI_2 growth in gels of different density (Halberstadt *et al.*, 1969). However, far too little is known about gel structures to permit firm conclusions. In many cases (though regrettably not all) greater gel density which intuitively implies smaller pore size does indeed diminish nucleation. On the other hand, it tends to increase the contamination by silicon and, thereby, to spoil the perfection and shape of the crystals. The density used in practice is therefore a compromise.

Another method of nucleation control involves the deliberate addition of foreign elements and is, of course, open to the same objections. As eqn (4.1.5) shows, the energy required for the formation of a critical nucleus involves Σ^3, where Σ is the surface energy per unit area, which is closely related to the surface tension (per unit length). Because of the third power, the nucleation probability is expected to be very sensitive to small changes of Σ and this, in turn, is sensitive to contamination, as surface tension always is. Moreover, an analogy with the process of solution is appropriate. For what are probably quite similar reasons, the solution process can be influenced by contaminants in the low parts-per-million range, as Ives and Plewes (1965) have demonstrated. As far as crystal growth in gels is concerned, corresponding experiments have been performed only on the calcium tartrate system (Dennis and Henisch, 1967), for which iron is the most effective contaminant. Fig. 4.4.1 shows the results of these experiments for the nucleation probability and contamination. The contamination increases monotonously, as one would expect, and in the course of it the crystals become yellow. (Crystals grown in commercial waterglass are often slightly yellow for the same reason.) Small ferric ion contaminations increase the nucleation probability, whereas larger concentrations inhibit it, with obvious penalties, however. The reasons for the reversal of the trend are not yet understood.

A control procedure which is free from the above objections involves concentration programing (Dennis and Henisch, 1967). There is, after all, no reason for believing that a reagent concentration which is optimum for nucleation is automatically optimum for growth. In the course of the procedure here envisaged, the concentration of the diffusing reagent is initially kept below the level at which nucleation is known to occur. It is then increased in a series of small steps, which can be optimized for

Fig. 4.4.1. Effect of iron in the gel on the nucleation and contamination of calcium tartrate crystals. After Dennis and Henisch (1967).

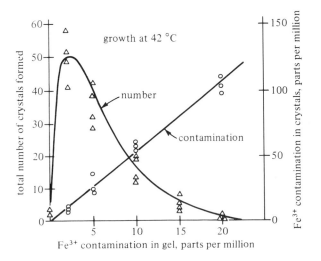

Fe^{3+} contamination in gel, parts per million

any system as regards magnitude and timing. At some stage, as the concentration of the diffusant increases, a few nuclei begin to form. It is plausible to assume that these act as sinks and result in the establishment of radial diffusion patterns which actually reduce the reagent concentrations in some of the neighboring regions. In this way, the formation of additional nuclei would be inhibited. Subsequent increases of reagent concentration lead to faster growth but not, in general, to new nucleation. The existing crystals are thus able to grow non-competitively, and their quality is correspondingly good. It has been found empirically that frequent small steps are more beneficial than a few large concentration increases.

Concentration programming has been successfully applied to the control of nucleation and growth in several systems (see Table 4.4.1), and has yielded crystals of larger size and a higher degree of perfection than those produced without programming. The size increase amounts to approximately a factor of 3, except for calcium tartrate, for which nucleation is rare enough even without programming to reduce the importance of competition during growth. Fig. 1.3.3(b) shows a PbI_2 crystal grown by concentration programming, in comparison with specimens (Fig. 1.3.3(a)) grown by the ordinary method. It remains to be seen whether the method can be more generally applied, the present indications being that it can. It has, for instance, been used successfully for the nucleation control of SbSI (antimony sulfur-iodide), formed by the reaction of SbI_3 (in HI) and Na_2S (in H_2O). The two reagents are diffused into a U-tube containing an HI gel. Small crystals of BiSI and SbSCl can be grown in analogous ways. Patel and Rao (1978) used the

Table 4.4.1. *Concentration programming (typical effects of one form of procedure on crystal size)*

Crystal	Typical sizes: largest linear dimensions, mm	
	Without programming	With programming
Calcium tartrate	~12	~12
Cuprous tartrate	1	3
Lead iodide	3	15
Thallium iodide	0.5	1.5
Calcium carbonate (aragonite spherulites)	0.5	1.5
Cadmium oxalate	2	5

same method to achieve crystals of $KClO_4$ substantially larger than those previously obtained by simpler ('static') methods.

In principle, one might also attempt to control nucleation by controlling the gel structure, but very little has been done along these lines. Because gels are neither liquid nor solid, there is a great shortage of methods for quantitative structural investigations. As a result, we have all too little information about the relationship between gel structure and nucleation probability. In an attempt to explore these matters, gels have been dialyzed to free them from unwanted reagents, and then rapidly frozen and vacuum-dried. By all outward appearances, the gels were not damaged by this procedure, and the silica structure remained essentially intact. Freeze-dried specimens were shadowed with 100–200 Å of gold, and inspected with a scanning electron microscope. Freezing at liquid nitrogen, dry ice and ice temperatures gave similar results. Typical structures as reported by Halberstadt and co-workers (1969) can be seen in Figs. 3.5.3 and 4.4.2. The gel is evidently not a simple three-dimensional silica network, as is sometimes supposed. It actually consists of sheet-like structures of varying degrees of surface roughness and porosity, forming interconnected cells. The cell walls are ordinarily curved. The differences between old and new gels were too small to show up in these experiments, but the differences between dense and light gels are noticeable (Figs. 4.4.2.(a) and (b)).

From pictures of this kind, one can estimate pore sizes, cell dimensions, and cell wall thicknesses. The cell walls seen in dense gels have pores from less that 0.1–0.5 μm compared with values from less than 0.1–4 μm in low density gels. The cell walls are thicker for the dense gels, but it appears that the cell size does not depend at all sensitively on gel density. It should be noted that the average pore sizes derived from electron microscopy need not correspond to estimates derived from diffusion experiments, since cell-to-cell diffusion is limited essentially by the pores of smallest diameter encountered along the diffusion path. Figs. 3.5.3 and 4.4.2 are typical of acid gels (initial pH5). As one would expect, the pH during gelling has a profound effect on gel structure. As the pH increases, the gel structure changes from a box-like network to a structure consisting of loosely bound platelets which appear to lack cross-linkages; the cellular nature becomes less distinct (Fig. 4.4.3). Plank and Drake (1947a, b) have shown that the pore structures of silica–alumina gels can be controlled, to a degree, by adjusting the aluminium content. Similarly, the cell size in gelatin can be diminished by cross-linking with formaldehyde but, as far as is known, such 'adjusted' gels have not yet been used systematically in the course of crystal growth optimization.

Fig. 4.4.2. Structure of dense and light (acid) silica gels (×1500). (*a*) 0.4 M Na$_2$SiO$_3$ in gelling mixture. (*b*) 0.2 M Na$_2$SiO$_3$ in gelling mixture. After Halberstadt *et al.* (1969).

(*a*)

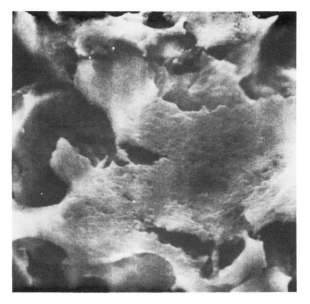

(*b*)

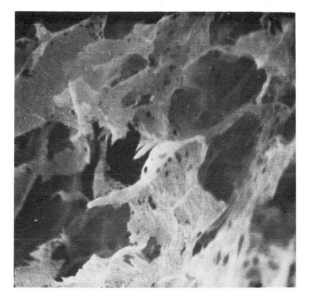

It is also of interest to compare silica gels with other gels (e.g. gelatin) in which crystals are known to grow. Fig. 4.4.4 shows that the cell walls of gelatin are smooth and relatively free of pores. The cells in gelatin are one order of magnitude larger than those in silica gel, and this is in harmony with the observation that gelatin supports much more nucleation.

Fig. 4.4.3. Structure of an alkaline silica gel (×2650). After Halberstadt *et al.* (1969).

Fig. 4.4.4. Structure of gelatin gel (×265). After Halberstadt *et al.* (1969).

4.5   Nucleation and crystal perfection

Although only one instance is described in Section 3.1, it is a common observation, made on a great variety of gel growth systems, that the crystals become increasingly scarce and more perfect with increasing distance from the gel interface. Several mechanisms are believed to be at work, singly or jointly, to bring this about.

(*a*) In cases involving the salts of weak acids, the environment becomes increasingly acid during growth, and the likelihood of a nucleus reaching critical size is correspondingly reduced. It has been shown (Halberstadt and Henisch, 1968) that calcium tartrate crystals do not ordinarily nucleate at pH values below about 3. Higher perfection at increasing depths would be the natural outcome of diminished competition. By adding hydrochloric acid to the diffusing calcium chloride solution, fewer and better crystals are produced nearer to the gel interface, which is consistent with the above conclusion. For the reasons given in Section 3.2, similar (but not identical) considerations apply to the growth of lead iodide, even though no strong acid is involved.

(*b*) At larger distances from the gel interface, where the number of crystals growing diminishes, the prevailing diffusion gradients are smaller, and this is believed to be the principal explanation of the relative crystal scarcity. However, there is a contributory cause arising from gel ageing (Halberstadt *et al.*, 1969). Crystals which nucleate in the lower regions obviously nucleate in an older gel. The existence of such a contribution has been substantiated by systematic experiments with gels after varying amounts of preageing. The results for calcium tartrate and silver iodide are shown in Fig. 4.5.1. There is the usual statistical spread, but the trend, though not actually spectacular, is significant enough. As one would expect in the circumstances, this trend is more pronounced for crystals which nucleate sparingly than for those which nucleate copiously. In lead iodide systems, which come into the latter category, no definite ageing density is observed. In contrast, ageing effects are particularly pronounced for the thin, fragile lead dendrites which may be grown in gels from lead salts on replacement of the lead by zinc. Faust (1968) showed that growth rates are drastically reduced with increasing time after gelling.

One possible explanation of the ageing effects might be that there is a progressive formation of cross-linkages between siloxane chains, resulting in a gradually diminishing gel pore size. This, in turn, should lead to a lowering of the nucleation probability, since many potential nuclei, whether homogeneous or heterogeneous, should find themselves in cells of too small a size (or in insufficient communication with other cells) to support growth to visible crystal sizes. By affecting the pore size distribution, an increased number of cross-linkages would also be expected to diminish growth rates.

Gel density should have a similar effect, and in this way the hypothesis can be checked. Fig. 4.5.2 gives a comparison of growth results obtained under otherwise identical conditions (pH and reagent concentration). There is no direct evidence that the mechanism whereby the nucleation rate is originally suppressed by the gel is the same as that active during ageing, but it is plausible to conclude that it is (Halberstadt *et al.*, 1969).

(*c*) Crystals at substantial depths in a growth system grow more slowly than those near the top, because of the smaller concentration gradients. There is also every reason to believe that the rate of solute arrival

Fig. 4.5.1. Effect of gel ageing on the nucleation of (*a*) silver iodide and (*b*) calcium tartrate; number of crystals growing more than 1 cm from gel surface. After Halberstadt *et al.*, (1969).

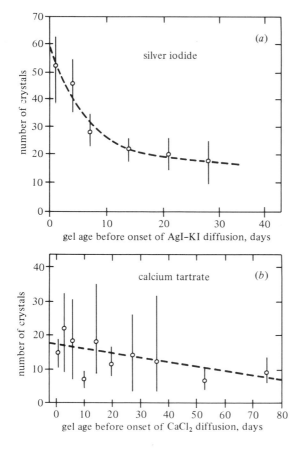

influences the perfection of the nuclei formed. A macroscopic crystal may be expected to maintain ϕ_0 constant during growth (see Section 3.1), but the same expectation cannot possibly apply to the nucleus. One could, alternatively, discuss the matter in terms of the instructive (albeit greatly oversimplified) eqn (4.1.8). Disorder among the first few constituents of a potential nucleus would on general grounds translate into a diminished value of Σ, and that, by reducing E_c, would increase the chances of the nucleus reaching critical size; hence more nucleation near the top. Unlike (a) above, this mechanism accounts for the crystal distribution without requiring any particular pH change. It also offers an alternative or supplementary interpretation of the success achieved by concentration programming previously described (Henisch *et al.*, 1965).

Fig. 4.5.2. Effect of silica gel density on the nucleation of lead iodide. (a) 0.1 M Na_2SiO_3 in gelling mixture. (b) 0.3 M Na_2SiO_3 in gelling mixture. After Halberstadt *et al.* (1969).

(a)   (b)

Investigations concerning the effect of light on the formation of crystals have an astonishingly long history: Kasatkin (1966) quotes references which go back to 1722, and that should satisfy the most fastidious researcher with baroque tastes. However, because conditions during nucleation and growth are not always (!) kept constant with the necessary degree of precision, the results obtained by different workers are often in conflict. Kasatkin reopened the whole question with reference to the growth of $NaBrO_3$ from solution, and established a positive effect beyond reasonable doubt. Light increased the growth rate, perhaps through the lowering of a potential barrier at the growth face, but possibly for more mundane reasons. Plausible models for such an effect should not be hard to find but, in the absence of more detailed data, their formulation appears premature.

The first reports of light effects on nucleation (as opposed to growth) in solutions date from 1900, and similar effects in gels have also been reported from time to time (Henisch *et al.*, 1965, Dennis, 1967). For instance, calcium tartrate crystals often grow more prolifically under, or after, irradiation of the gel system than in darkness. It is tempting to envisage a new and intriguing quantum process ('photonucleation'), but the available observations lack the necessary generality, and do not lend themselves to an interpretation along these lines. In any event, it was soon found that irradiation of the gel before the diffusion process (followed by nucleation in darkness) produced very similar results. That effect was traced to the light-stimulated development of small carbon dioxide bubbles, either by dissociation of some tartaric acid or by the release of some carbon dioxide previously dissolved. This may account for the lack of reports to date on light-stimulated nucleation in connection with simpler inorganic gel systems. On the other hand, strong evidence for a genuine light effect on epitaxial growth (and thus heterogeneous nucleation) has been reported by Kumagawa and co-workers (1968) in the course of vapor growth experiments, and it is, of course, quite plausible that similar processes might be at work in gels. One problem is to isolate them, another is to cope with nature's fickleness. Thus, for instance, Armington and co-workers (1967) found that illumination had an adverse influence on the perfection of gel-grown cuprous chloride crystals.

Light can alter the distribution of impurities between the growing crystal and the surrounding gel (Henisch *et al.*, 1965) and is may well be a surface barrier effect as suggested above. This is easily demonstrated with calcium tartrate systems in the presence of iron contamination. Under irradiation by ultraviolet or visible light, the iron uptake is inhibited, and the resulting crystals are almost colorless; those grown in darkness are yellowish. These matters deserve a great deal of further investigation.

Among the researchers who have described light effects on Liesegang Ring formation are Hatschek (1921, 1925), Miyamoto (1937), Köhn and Mainzhausen (1937) and Roy (1931); see also Section 5.2.

5

Liesegang Rings

Crystal growth in gels has been a subject of intense interest for many years, but the process has remained a laboratory research activity. Certainly, it has not (or, shall we delicately say, not yet) given rise to a new industry, but it is nevertheless recognized as useful. In contrast, it would until quite recently have been necessary to admit that Liesegang Ring formation, for all its aesthetic merits, has yet to find its practical niche. One would have had to argue (as, indeed, one still does) that the study of this beautiful phenomenon is instructive in a general way, and it is its own justification. However, a new possibility has emerged which changes this position, namely that of designing anisotropic ceramics with Liesegang microstructures. This is a potentially potent technology but, of course, outside the immediate context of 'crystal growth in gels.'

5.1.   R. E. Liesegang

A tribute, first, to the discoverer who gave this set of phenomena his name. Somehow the name Liesegang has an ancient ring, and it is tempting to believe that its owner was one of the earliest chemical pioneers. In fact, Raphael Eduard Liesegang (born 1869) was as much a physicist as he was a chemist, though it was never easy to classify him into any professional niche. He lived until 1947, and did many things in his lifetime that would make a modern scientist recoil in awe. Thus, for instance, he wrote one of the earliest papers on the possibility of television (in 1891, and before the discovery of electrons), was active as a bacteriologist, contributed to chromosome theory and to the beginnings of paper chromatography, not forgetting the properties of aerosols and gelatins, the origins of silicosis, the mechanism of the photographic process in black-and-white and color (1889). Somewhere inbetween he also managed to publish a paper on the role of carbon dioxide in plant life.

When he was 29 years old, he published his famous book on chemical reaction in gels (1898), a topic to which he was drawn by his even earlier interest in photography (see also Liesegang, 1896). This interest was in the blood, so to speak, since J. P. E. Liesegang, his father, and F. W. E. Liesegang, his grandfather had both been pioneers of photography and the photographic industry (Huttel, 1984).

This is clearly not the summary of an average scientist's career; indeed, the very notion of being 'average' in any sense would have revolted Liesegang, as his portrait (Fig. 5.1.1), taken in 1938 or 1939, convincingly suggests. On the occasion of Liesegang's 70th birthday, his friends and colleagues issued a Liesegang *Festschrift* (Ostwald, 1939) under the editorship of Wo. Ostwald, always so identified to distinguish him from his namesake and senior Wi. Ostwald, who was also active in the field. This volume likewise reflects the great variety of his interests. Ostwald

Fig. 5.1.1. R. E. Liesegang at the age of 70. From Ostwald (1939)

wrote the introduction and refers to Liesegang's 'incredible universality', and to the difficulties of providing an adequate overview of such a man's work. He had, indeed, been stunned by that task on a similar occasion ten years earlier and had then, for want of a better solution, asked Liesegang for a self-assessment, which R.E.L. duly provided 'to prevent other people from spreading false notions about me'. That autobiography, says Ostwald, is itself a typical piece of Liesegang, and it really is.

'I created a commotion from my first schooldays onward', he begins with unassuming modesty, and goes on to explain that he was always a stubborn and unresponsive pupil, as long as he found himself in situations in which learning was an imposed obligation. As the son of a painter (his father's original profession), he longed for an artistic career, but had no talent in that direction. As he put it, not even his name Raphael did him any good. In any event, a new camera invented by his father 'made drawing unnecessary', and though that was a modern instrument, young Liesegang delved into photographic history by using the lid of a silver watch he had been given as a confirmation present for experiments on the daguerreotype. Photography continued to hold his attention with a strength sufficient to make him a failure at school, but his interests broadened later, and no one can say that he failed to make up for lost time.

Liesegang's first published paper dealt with the light sensitivity of potassium iodide. He received an honorarium of 20 marks for it, with which he did not know what to do, because, as he said: 'I detested beer and had not yet begun to smoke'.

At the University of Freiburg he attended only one lecture (on botany), finding the life in coffee houses much more interesting, and also the company of theater people which he began to cultivate. Even laboratory classes held no attraction for him, but he did study experimental psychology for a semester. In the circumstances, he decided to abandon all thought of taking final examinations in anything, yet another display of good judgment.

From 1892 onward, Liesegang worked for his father's photographic factory, experimenting with light-sensitive emulsions, and when the job demanded that he keep his findings confidential, he published them anyway under a series of pen names. A little later he was entrusted with the editorship of the *Photographisches Archiv*; he mentions it in passing, without explaining how such an, on the face of it, unlikely thing came about. Morse and Pierce (1903) refer to Liesegang as 'the German photographer'; it was his technical interest in photographic emulsions that actually led him to research on gels; see Liesegang (1896, 1897).

When Liesegang senior died in 1896, Raphael Eduard took over and thoroughly modernized the factory by introducing the latest machinery. His hope at the time was that he would eventually overcome his prejudices against the life of commerce, but that never came to pass, and so he was greatly relieved when another manufacturer expressed an interest in purchasing a large part of the company. In 1908, he published his *Contributions to the Colloid Chemistry of Living Matter*, and in 1914 he spent a period of research in a neurological institute. 'By the side', he says, 'there came about' some publications on the structure of agate. During much of World War I he was director of a anaesthetics division of the army's First Aid depot at Frankfurt. Even there, he did some research, and another publication 'arose by the side', this time on the toxidermy of wound plasters.

The great inflation that hit Germany after World War I made it necessary for him to work 'for money', first as a consultant to a company that made photographic paper, and later for a pharmaceutical company as well. He performed 'over 10 000 experiments' in that capacity, alone, without ever having an assistant by his side. 'My father's life-long dislike of machinery struck home; I did not want any apparatus. A couple of glass plates, some test tubes and a few dishes were all I needed. I am glad I never had to teach; it allowed me to remain a student all my life'.

5.2   Qualitative features: spiral formation; radiation effects

A hint of the manifold appearances of 'Liesegang Rings' has already been given in Fig. 1.2.1(a)–(c), sufficient to show that problems of substantial complexity are likely to be involved in the formulation of any convincing theory. Thus, the rings can be microcrystalline, bordering on amorphous, as in (b), or microcrystalline, bordering on macrocrystalline, as in (c), or distinctly macrocrystalline, as in (a). The bands can be few or many, broad or narrow, obeying simple spacing rules, as in (a), or obeying them in part, as in (b), or in no obvious way at all, as in (c). Nor does this exhaust the variety of observed configurations. Thus, for instance, Buchholcz (1941) has reported on the formation of well-defined layers (as distinct from separate rings) of different colors and transparencies in iodine-containing gelatin. To complicate matters, irradiation by visible light was found to suppress these layers, and to encourage the development of 'normal' Liesegang Rings.

Ring formations in great variety have been described by Stansfield (1917), Orlovski (1926), Kuzmenko (1928), and Daus and Tower (1929). Among the most exotic structures reported are bands of chloroform made by diffusing potassium hydroxide into agar gel containing chloral hydrate

(Füchtbauer, 1904), and bands of metallic mercury, made by diffusing mercurous nitrate into an agar gel containing sodium formate (Davies, 1917), see also Knöll (1938*a*, *b*) for rings of urea. Because particle size can affect color, ring systems are sometimes multicolored. Hatschek (1919) described such a case, involving cadmium sulfide precipitates. Buchholcz (1941) concluded that 'many factors must be involved', and that 'the situation is not yet fully clarified', a summary which has stood the test of time.

When Liesegang Rings are grown in tubular structures, as so far envisaged, they are, of course, not 'rings' at all, but disks. However, the periodic deposition does appear in the form of concentric circular rings, when one of the diffusing reagents spreads radially in a plane, e.g. as shown in Fig. 5.2.1, and it is in this form that Liesegang originally discovered the phenomenon. He used gelatin layers set on glass plates. The gelatin contained ammonium bichromate and a little citric acid. A drop of silver nitrate, placed at the center, served as the source of silver, albeit as a highly non-constant one. A ring system took between two and four days to grow, and Liesegang (1914) reported that 'a cool room is better than a hot one'. Distressingly, he also found that unpurified, commercial gelatin worked better than the purified product. Among Liesegang's interesting observations was that the ring spacing appeared to be independent of the silver concentration, but though this may have been true for his systems, it did not establish a general rule. Liesegang himself concluded that the effect of additional silver was counter-acted by the additional nitric acid produced as a waste-product of the reaction.

Rings of this general type have been observed in great variety, and their analyses, as well as the analyses of the now more often studied disks, must evidently proceed along similar lines. There are, however, findings on record, first made by Rothmund (1907), which differ markedly from the familiar pattern described above though, as a matter of fact, Rothmund never realized this himself. He had placed a drop of silver nitrate at the center of a gelatin-filled dish, the gel itself containing chromium ions, and he observed a fine system of what he considered to be concentric, circular rings. Liesegang (1939) subjected the Rothmund pictures to more detailed scrutiny, and found that they did not show circles at all, but concentric spirals. The example in Fig. 5.2.2(*a*) is a six-membered spiral, and Liesegang (1914) had already observed on an earlier occasion that the various components (whether spiral members or rings) are not generally in the same plane. He ascribed this separation to an inhomogeneous (layered) structure of the gelatin.

In other cases still, systems were found to consist of circles, as long as the radii were small; they changed later into spirals as the radii grew larger. Liesegang (Fig. 5.2.2(*b*)) even found a bifurcation, but pointed out that the growth does not in fact (ever) proceed tangentially (i.e. along) the spiral. Instead, different parts of the spiral make their appearance at unrelated times, with fragments joining later. He also mentions that several earlier workers, Tillians and Heublin (1915), Hatschek (1920, 1921, 1925), Franz (1910), and Gebhardt (1912), as well as Ostwald (for

Fig. 5.2.1. Details of concentric ring system in agar. After Hatschek (1914). Janek (1923) likewise reported observing such rings, without, apparently, being aware of earlier work in the field. (*a*) Silver chromate (wet) × 100. (*b*) Lead iodide (wet) × 100. (*c*) Lead iodide after drying × 50. (*d*) Lead chromate (wet) × 50.

(*a*)

(*b*)

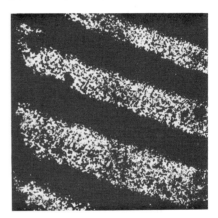

(*c*)

(*d*)

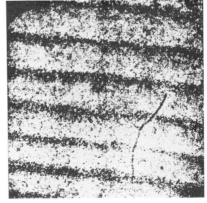

whom he gives an erroneous reference), had actually observed three-dimensional spirals in cylindrical (as distinct from planar) growth systems. Hatschek (1925) gives a particularly beautiful example of a three-dimensional staircase. Liesegang struggles with the problem, but turns

Fig. 5.2.2. Silver chromate spirals in gelatin. (*a*) Six-fold spiral, after Rothmund (1907), reproduced and discussed by Liesegang (1939), who added the dots manually in order to make the spiral character of the lines more obvious. (*b*) Spiral with a bifurcation, after Liesegang (1939). (*c*) Heavy spiral, after Liesegang (1924).

(*a*)

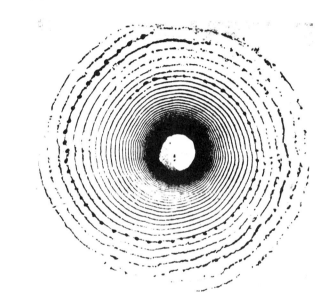

(*b*)

(*c*)

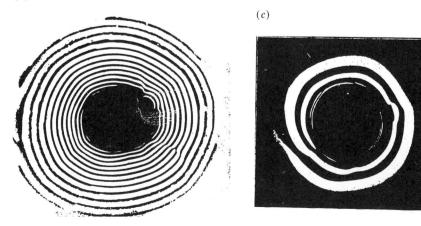

it aside without offering a clear-cut solution. He must have been aware
of the fact that true spiral formation would involve something very
profound, i.e. a rotating vector of one kind or another, whereas no
mechanism of this kind was anywhere in sight. Liesegang therefore
preferred to believe that the observed spirals were an artifact, 'arising
from small and accidental disturbances', disturbances, that is to say,
imposed on an essentially concentric ring system. The hypothesis of
non-uniform gel properties could not by itself provide the answer, since
it would imply merely a system of continuous, concentric, but non-
circular contours. However, as Dhar and Chatterji (1925a, b) have
pointed out, none of these ring systems is actually static. Rings form and
dissolve, are replaced and dissolve again. Such changes, coupled with a
non-homogeneous, layered gel structure in which rings form and move
at different levels could indeed lead to the occasional illusion of spiral
growth, and that is (without explicitly saying so) what Liesegang must
have meant when he dismissed the phenomenon as unimportant. In this
connection it is also interesting to recall the work of Burton and Bell
(1921) who performed (rather informal) experiments on gels made
anisotropic by stretching. Obrist (1937) seems to have been the first to
draw attention to a rather obvious fact, namely that rings, as traditionally
grown in a plane and exposed to air, *dry* in the course of time, and that
the drying process leads to all kinds of mechanical distortion; see also
Popp (1925) and Lloyd and Moravek (1928) for references to 'Saturn
structures', arising (in a complicated and as yet very little understood
way) from the disturbing influence of internal growth tube surfaces. It
may be that the axial feature in Fig. 1.2.1(*b*) has such an origin.

Among the interesting observations of 'exotic' ring phenomena are
probably those of Müller *et al.* (1982b) who reported almost circular
PbI_2 rings with distinct perimeter gaps, at times combining to form radial
features. They ascribe the process of formation to competitive particle
growth (see Sections 5.4 and 5.6), and this is undoubtedly correct.

Liesegang (1937) found that a spreading ring system may encounter
a new reagent (whether deliberately added or accidentally present in the
gel) and thereupon change the entire pattern of deposition; he called
such agents 'form catalysts'. In general, Liesegang maintained that the
most basic processes involved in spiral formation are the same as those
involved in ring formation, and this is bound to be true.

To the complexities so far described one should certainly add the fact
that something like Liesegang Ring formation is sometimes observed in
bacterial growth, e.g. as described by Knöll (1939) and Schmidt (1939).
This growth could be thought of primarily as a consequence of nutrient

diffusion and local exhaustion, but the differences between it and the more familiar inorganic growth may not be highly significant. Bacteria would first spread in search of nutrient contained in the (agar) medium, and would consume that nutrient not only within the circle of residence but, by creating a diffusion gradient, within a ring of a certain thickness beyond it. Within that ring, bacteria could not subsequently flourish, whereas they could flourish beyond it, having spread in search of food and reached that area. The process would then repeat itself. Fig. 5.2.3 shows a colony of *bacterium vulgare* (proteus). The rings denote different thicknesses of population, and hence different transparencies. In Fig. 5.2.3 they are highly non-circular, but much more regular growths have also been observed. In these systems, the bacteria produce a local exhaus-

Fig. 5.2.3. Liesegang Ring colony of bacterium vulgare (proteus) in agar. After Knöll (1939).

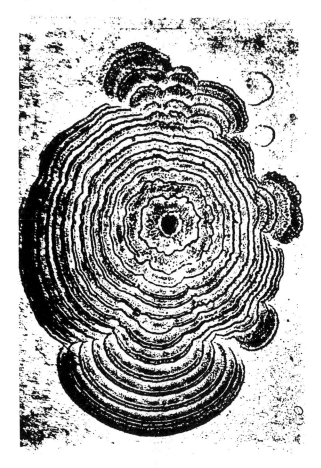

tion of nutrient by eating it; in the more conventional systems the growing crystallites 'eat it', and that situation is not so very different; see also Holmes (1918).

Liesegang Rings actually consisting of organic precipitates (urea, for instance) have also been widely studied, e.g. by Knöll (1938a, b), who gave many references to previous work. One of the peculiarities reported by Knöll is the great rapidity of ring formation and movement in his systems, with clearly visible changes taking place in a matter of minutes. See Koenig (1920), Karrer (1922), Copisarow (1931) and Hedges (1932) for descriptions of ring systems arising from gaseous reagents with chemically charged gels. These methods may lend themselves to single crystal growth under pressure.

Among the more exotic (but, in fact, not all that unexpected) observations is one by Gerrard *et al.* (1962) on the formation of AlAs rings arising from counter-diffusion in solid copper. These rings have been beautifully documented, see Fig. 5.2.4. Ring-free regions near the aluminium source were solid solutions of aluminium in copper, those near the arsenic source solid solutions of arsenic. Likewise in the category of exotica are rings formed in air by gaseous diffusion, as already mentioned in Section 3.2.

In view of the variety and complexity of the observed ring phenomena, it is not at all surprising that many different hypotheses have been put forward in attempts to explain them, and quite a number have received limited confirmation. Others have not been verified in any specific way. Thus, typically, Köhler (1916) maintained that the diffusion constants must be in a certain relationship to the 'crystallization speed' of the

Fig. 5.2.4. Photomicrographs of copper, with rings of AlAs resulting from the counter-diffusion of Al and As. After Gerrard *et al.* (1962). Contributed by F. R. Meeks.

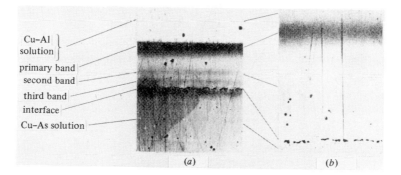

reaction products, if distinct rings are to be obtained, and this might well be true, but many other factors are evidently at work; see also Section 5.6. One of the greatest problems is that much experimental work has been performed by non-standardized (and often downright sloppy) methods, with non-standardized gels and non-standardized reagents, making it very difficult to compare one observation with another. An attempt at procedural improvement and standardization was made by Obrist (1937), who pointed out that gel spread over a glass plate does not remain constant over the period of experimentation. He also drew attention to the fact that radial diffusion of a reagent from a central source in the form of the conventional 'drop' implies a highly non-constant source. To overcome both problems, he kept his gels between two glass plates of which the upper was fitted with a tubular reservoir at the center, as shown in Fig. 5.2.5(a). It will be shown in Section 5.6 that a non-constant reservoir is not actually an impediment to ring formation, and in some ways, indeed, the reverse, but, however that might be, Obrist was rewarded with ring systems of sensational crispness and circularity, e.g. as shown in Fig. 5.2.5(b). It should be mentioned that the medium itself can, in certain circumstances, undergo periodic changes of structure; Moeller (1916, 1917) has described such cases, involving gelatin.

An experiment of great simplicity and potentially great significance for the interpretation of Liesegang phenomena was performed by Flicker and Ross (1974). The authors grew PbI_2 ring systems in agar gels. In a typical case, a 0.5% agar gel was charged with 0.3% KI and 0.15% $Pb(NO_3)_2$. Such a mixture is homogeneous, but far from equilibrium and therefore metastable. Supernatant solutions (0.03%, 1.15%, 1.3%) of KI were placed on top of the gel, and the systems were kept undisturbed at room temperature for several days. Some banding developed in all of them, even in those in which the KI concentration was higher in the agar than in the supernatant liquid. In such cases, KI should diffuse *out* of the gel, and the saturation level should actually *diminish*, rather than increase. The question is, what do the results signify? The authors interpreted the formation of rings under these conditions to mean that supersaturation can be ruled out as the driving force, but though the results suggest this, they do not incontrovertibly prove it. One could alternatively envisage that some KI is adsorbed on agar cell walls, thereby creating a distinction between the total and the free KI content of the gel. In that situation, one could still foresee KI diffusion from the supernatant liquid *into* the gel, leading to supersaturation. The matter deserves to be further investigated.

Later work in the same laboratory (Kai *et al.*, 1983), also on PbI_2 (grown in agar by counter-diffusing KI and $Pb(NO_3)_2$), revealed another intriguing fact. The authors examined ring formation for various concentrations A_G and B_R. A single first sharp band was formed with $B_R = A_G$, but 'two clearly separated sharp bands' appeared as $B_R - A_G$ increased, while $A_G B_R$, was kept constant, and, indeed, additional rings as $B_R - A_G$ increased further. Fig. 5.2.6 shows this. It also shows that the bands are located within a broad zone of low density colloidal PbI_2; see also Section 5.3. For $B_R - A_G$ large, these and later bands appeared in well-defined positions; for $B_R \rightarrow A_G$, the band positions became increasingly probabilistic, and thus less reproducible. For other observations

Fig. 5.2.5. Liesegang Rings of $Mg(OH)_2$ as grown in gelatin by Obrist (1937). (*a*) Obrist's method. (*b*) Example of ring system. Plate spacing: 0.5 mm. Plate size: 14×14 cm The gelatine contained $MgCl_2$, into which a solution of NH_3 was diffused. Growth time: 7 days.

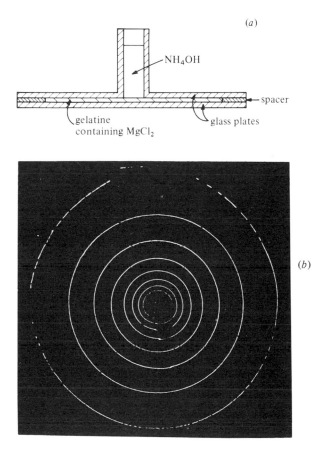

along similar lines, see Müller *et al.* (1982a). It would be interesting to see whether such results can be reproduced by computational methods of the kind explored in Section 5.6.

The same authors (Kai *et al.*, 1982) are also on record with what is believed to be the only set of experiments designed to test for the effect of gravity. Again, PbI_2 was grown in agar, but in this case the reagent reservoir was also gelled; see Section 2.3. Three growth conditions were examined, (see Fig. 5.2.7) and the results can be summarized as follows:

   (*a*) KI above $Pb(NO_3)_2$; growth downward; $\Delta y/y$ large.

   (*b*) KI at same level as $Pb(NO_3)_2$; growth horizontal; $\Delta y/y$ smaller.

   (*c*) KI below $Pb(NO_3)_2$; growth upward; $\Delta y/y$ smallest.

With convection currents ruled out, the explanation must obviously lie in the effect of gravity on such residual Brownian movement as the gel structure (depending on its density) allows. Such phenomena are also believed to play a role in some sedimentation processes.

In view of the gross lack of standardization, the episodic history of early experimentation does not need to be exhaustively pursued at this stage. For a discussion of early work, see, for instance, Dhar and Chatterji

Fig. 5.2.6. Liesegang Ring systems of lead iodide, for different values of $\Delta = B_R - A_G$, the values of Δ for tubes 1–13 being respectively 2.6, 4.2, 5.6, 7.1, 8.4, 10.8, 12.1, 14.0, 16.1, 18.2, 20.0, 22.6, and 25.0 mM. $B_R A_G$ was kept constant. Ring formation shown as recorded after 6 days. After Kai *et al.* (1983). Contributed by J. Ross.

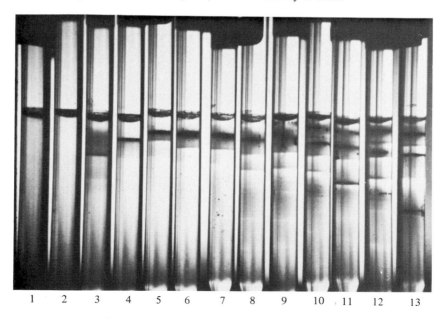

  1   2   3   4   5   6   7   8   9   10 11   12   13

(1924, 1925a, b) and the many references given there, and also Stern's (1954) extensive overview. An extensive bibliography by Stern (1967) contains 786 items. A book by Hedges on the earliest work appeared in 1932. It is, however, desirable at this point to return to the subject of radiation. The effect of light on single crystal growth has already been

Fig. 5.2.7. Effect of gravity on Liesegang Ring formation. Both reagent reservoirs gelled. (*a*) KI above PB(NO$_3$)$_2$; growth downward; $\Delta y/y$ large. (*b*) KI at same level as Pb(NO$_3$)$_2$; horizontal growth; $\Delta y/y$ smaller. (*c*)KI below Pb(NO$_3$)$_2$; growth upward; $\Delta y/y$ smallest. After Kai *et al.* (1982). Contributed by J. Ross.

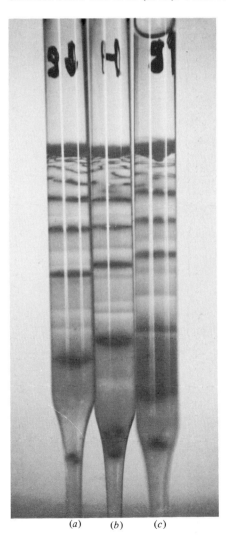

(*a*)   (*b*)   (*c*)

referred to in Section 4.5, but much more work has in fact been done on Liesegang Rings. Hatschek (1921) discovered the light effect accidentally, and reported that the ring spacings in his lead chromate systems were different for formation in darkness and under light. He did not offer a firm explanation, but hypothesized that the hardening effect of light on organic media containing dichromate ions might have something to do with it. That effect was already well known (from its uses in photographic technology) but had to be dismissed as a possible explanation when independent tests failed to confirm it for agar (as distinct from

Fig. 5.2.8. Light effect on the formation of silver chromate rings in gelatin; the effect of chlorine content. D = dark, L = light After Köhn and Mainzhausen (1937). (*a*) Ring system without chlorine; no light effect. Magnification: × 1.5. (*b*) Ring system with 1/4000 M chlorine content; prominent light effect. Magnification: ×14.

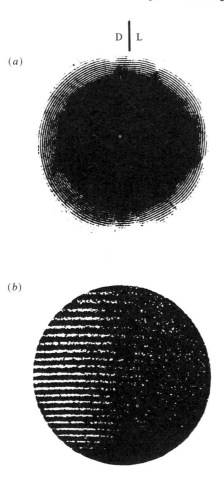

gelatin). Hatschek did not think in terms of nucleation and growth barriers, at that time, but in a later paper, see Hatschek (1925), he did suggest that light might affect the formation of crystallization nuclei.

Küster (1913) worked on silver chromate deposition and obtained no ring formation in his systems while these were kept in darkness, but prominent rings upon 'rhythmic' (i.e. intermittent, but otherwise unspecified) illumination. Dhar and Chatterji (1934) did obtain rings in darkness, but found their number increased by light, and the spacing between them diminished. Both Roy (1931) and Kisch (1933) reported that their rings were more distinct when formed 'under light' than in darkness. Köhn and Mainzhausen (1937) believed that light effects should be particularly strong in gelatin containing silver halides, and subjected such systems to more extensive examination. A set of gelatin specimens was prepared, containing the same amount of potassium chromate, but systematically varying amounts of chlorine. Half of each dish was kept in darkness, the other was illuminated. In the absence of chlorine there was no light effect (Fig. 5.2.8(a)); in the presence of chlorine the shadow edge was very distinct (Fig. 5.2.8(b)). A very small amount of chlorine (1/4000th M) was sufficient to demonstrate the effect. In contrast, Gnanam *et al.* (1980) found no light effect of any kind, on Liesegang Ring systems of calcium carbonate in agar.

Miyamoto (1937) performed similar experiments on silver chromate in the presence of an electric field (the relevance of which was not made totally clear). His observations were that (a) visible light had practically no effect, (b) ultraviolet light diminished the ring thickness, diminished the ring spacing and also led to colored rings, and (c) irradiation by ultraviolet light before diffusion had no effect (unlike the observations discussed in Section 4.5.).

From all this, we gather that the light effects are very real, and that the field is greatly in need of reexamination by methods more modern and sophisticated than those available to the pioneers quoted here.

5.3   Problems of quantification; general

Research on single crystal growth tends to have very practical objectives, mostly the production of specific crystals intended for other kinds of experimentation. In contrast, research on Liesegang Ring phenomena tends to be a labor of love, but it is none the worse for that.

The discussion has so far been concerned with qualitative features, and these have fascinated a good many researchers in their time, but the urge to penetrate beyond the descriptive stage into quantitative analysis has always been strong. Ideally, one would be looking for some expression which gives the density of deposited crystalline material at any point of the growth system as a function of time. It was always recognized that such an expression, if it existed, would have to be a function of many variables, e.g. the reservoir concentrations (whether constant or time-dependent), the solubility product (K_s), the precipitation product (K_s') and, as we saw in Section 4.1, some factor which depends on the concentration ratio of reacting components. The expression would also need to depend on the highly complex descriptive parameters of the gel (including gel density, pore size, and pore connectivity), the diffusion constants of the reacting species, and at least one term governing the adsorption of material on internal cell walls. Several parameters would be temperature-dependent and, moreover, dependent on the intensity and spectral composition of incident light.

Somewhere one would also have to find room for purely geometrical factors, and that should be the simplest matter of all. Far worse is to come: we must expect the descriptive parameters of the medium to be modified locally by ageing on the one hand and by the ring formation itself on the other. This means that the diffusion 'constants' would be time- and space-dependent. Thus, for instance, Lee and Meeks (1971) found that Liesegang Rings of lead bichromate in agar 'exhibit membrane-like characteristics by selectively accumulating potassium and calcium ions, but allowing sodium, magnesium and a wide number of other inorganic ions to diffuse through freely'. Such membrane characteristics were first reported by Hirsch-Ayalon (1957); see also Van Oss and Hirsch-Ayalon (1959).

In view of all this, it is abundantly clear that the search for a general expression as envisaged here would be 'a tall order'. Certainly, nothing like it has ever been achieved, nor is its achievement a reasonable expectation for the future. Meanwhile, one finds that a great deal of sophisticated but functionally empty mathematics has often been lavished on the problem. Ultimately, we must come to terms with the fact that the relationships involved do not *have* an analytic solution in a general form. This applies not only to the diffusion equations, but also to the equations which Van Hook (1940) used to describe the dynamics of precipitation and precipitate growth. Early analysts were not unaware of the difficulties, but proceeded nevertheless, in the hope that at least some systems would allow themselves to be simplified and idealized. Many such attempts are on record. It would have to be admitted that they are not ultimately successful, but two such treatments are nevertheless described (see Section 5.4), partly for historical interest, and partly because various aspects are instructive.

The alternative is to turn to the computer. That involves some problems of its own, but it certainly by-passes the lack of analytic solutions, and this is no minor advantage. A truly comprehensive computer analysis is not yet available, but substantial beginnings have been made (see Section 5.6) and have shown enough promise to make this the preferred approach. Odd as it may sound, the computer actually allows us a degree of insight that is not accessible to us (even) by experimentation, inasmuch as it allows parameters to be changed one at a time, and in total independence of all other parameters. No 'real-life' system allows such a thing.

There remains the general problem of comparing analysis (any analysis) with experiment. What should be measured, ring position or the time of first ring appearance? In the absence of standardized criteria, one must also ask: at what stage can a Liesegang Ring be diagnosed as duly 'formed'? Only then would it make sense to ask what this moment signifies in terms of microscopic processes. Morse and Pierce (1903) claimed to be able to judge the moment of first appearance to the nearest second, which is itself surprising. There is, after all, no simple answer to the question of what constitutes an 'observable' ring. How many particles of what size and distribution are needed for a deposit to be seen, and with what type of equipment? Laser ultramicroscopy, e.g. as used by Vand *et al.* (1966) and already mentioned in Chapter 3, would once again be a very suitable technique for investigating these matters and, even more, measurements of the intensity of scattered light from a laser beam, as used by Kai *et al.* (1982). These workers also employed optical transmission and the refractive index for diagnostic purposes. Overall AC conductance measurements on systems as a whole have been used to monitor precipitation (Lucchesi, 1956), but they are not by their nature capable of providing sufficiently local information to be of direct help.

It goes without saying that, for a Liesegang Ring to be visible to the naked eye, the mean distance between neighboring crystals must be smaller than the ring spacing. There are cases in which this is not an easy judgment to make, which means that there is not necessarily a sharp distinction between ring formation and continuous distribution. In any event, visibility by the naked eye is not a cogent criterion, since it is known that precipitates (presumably colloidal) are often formed which are *not* visible. Higuchi and Matuura (1962), as well as Fricke and Suwelack (1926) have reported such findings. Whether this is *always* true remains unresolved for the moment, but the fact that it happens at all should be enough to induce scepticism in even the staunchest enthusiast for ring distribution laws.

If time measurements are uncertain, distance measurements (though superficially simpler) have their own problems. For instance, on the basis of equations similar to those cited in the next section, Kahlweit (1962) made estimates of the ring spacing and, in agreement with earlier results by Wagner (1950), predicted that $\Delta Y / Y$ should be constant. At first glance, this seemed to offer a fine opportunity for a critical test, but one of the underlying assumptions was, of course, that $\Delta Y \ll Y$. In practice, therefore, such tests had to be performed far down the growth tube, where only a few rings were left. Nor, indeed, is it a highly critical test, since almost any profile that can be vaguely approximated as an exponential or power law over the observed interval will obey it. As an example of the latter, Matalon and Packter (1955) expressed the empirically determined spacing relationship as

$$Y_i = gh^i \tag{5.3.1}$$

where i denotes the ith ring, and Y_i the ith position; g and h are constants, the latter being called the 'spacing coefficient'. From this

$$Y_{i+1} - Y_i = \Delta Y = gh^i(h - 1) = Y_i(h - 1) \tag{5.3.2}$$

making $\Delta Y / Y$ constant. However, these and similar formulas convey an altogether excessive impression of precision and control, whereas the rings formed can actually vary from very crisp to barely visible; see Jablczynski (1923), Schleussner (1924), Lakhani and Mathur (1934), Serb-Serbina (1933), Holmes (1918), and Gnanam *et al.* (1980). Ramaiah (1939) reported seeing spacings of less than 0.01 cm, and such microstructures (in silica as well as alumina gels) have also been recorded by Adair (1986). Even in cases where the measurements themselves are simple (e.g. as described by Palaniandavar *et al.*, 1986) the task of investigating the dependence of the descriptive parameters and of generating stable conclusions is enormous.

A particularly crisp example of a ring structure is shown in Fig. 5.3.1(a). It consists of copper borate rings grown by diffusing $CuCl_2$ into a silica gel containing H_3BO_3 and took several months to form (Behm, 1984). Such configurations certainly make the exploration of spacing laws very tempting, but Fig. 5.3.1(b), observed in another such case, shows an unusual feature, one that can play havoc with spacing laws: a thinning of rings is followed by a thickening. It is as if two groups of rings were here superimposed, for reasons which are not yet clear. Such a formation, indeed an even more prominent one (originally due to Hatschek), was also shown by Liesegang (1924); see Fig. 5.3.2. 'Much is not yet understood', is what Liesegang said at the time, and this has remained true ever since.

One of the (by now) classic issues is presented by the occasional appearance of 'inverted rhythmic precipitation', as originally described by Packter (1955) for silver chromate systems 'in the presence of a third peptizing ion of slower rate of diffusion'. Whereas the initially formed rings obeyed eqn (5.3.1), rings formed at the lower end of the tube conformed to

$$Y_i = gh^{-i} \tag{5.3.3}$$

Packter's description of the growth method is not totally free from ambiguities, but such formations (often, but not elegantly, called 'revert' rings) have since been widely observed in other systems, see also Section 5.6, where a (computer-generated) case is cited that does *not* depend on

Fig. 5.3.1. Liesegang Rings of copper borate in silica gel, made by diffusing $CuCl_2$ into a gel containing H_3BO_3: (*a*) and (*b*) are for different $CuCl_2$ concentrations; note the secondary ring system in (*b*). Contributed by Behm (1984).

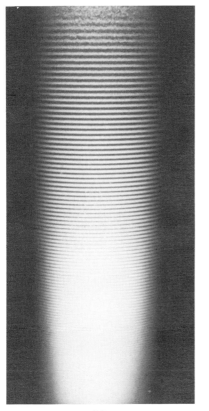 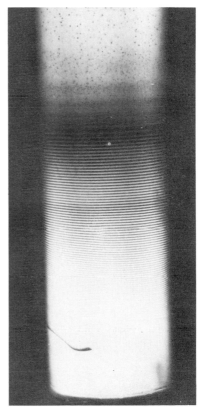

(*a*)   (*b*)

the presence of a third ion. Examples of 'revert' ring systems of lead chromate are shown in Fig. 5.3.3, and similar observations are on record for silver iodide, see also Kanniah *et al.* (1983), and Krishnan *et al.* (1982). Since most parameters of such systems are temperature-dependent, one would expect the spacing pattern to be temperature sensitive also; see Ambrose *et al.* (1984).

Adair's (1986) observations were made in a very different context, namely that of preparing gels and, ultimately, sintered ceramics with built-in Liesegang fine structures. Two examples of $CuCrO_4$ rings are shown in Fig. 5.3.4. It was also shown that these ring structures are at least partially conserved during sintering. In such silica gels, Adair obtained very pronounced differences in ring spacings, depending on whether Cu_2SO_4 or K_2CrO_4 was supernatant, and this strongly suggests that adsorption played a role, presumably on internal cell walls, but perhaps also on the precipitated particles themselves.

In connection with the characterization of ring systems, additional problems arise on the one hand from the tricky matter of ring *thickness* (which tends to make the ring position ill-defined) and on the other from the even trickier one of ring movement (see Section 5.5). As a result, the significance of many measurements quoted in the literature is seriously in doubt. If truly informative results are to be obtained, all these matters

Fig. 5.3.2. Liesegang Rings of lead chromate in agar, originally by Hatschek, quoted by Liesegang (1924). Note prominent secondary ring system.

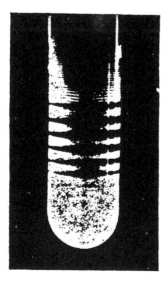

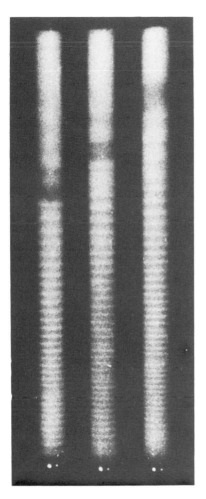

Fig. 5.3.3. 'Revert' Liesegang Rings of lead chromate in (1%) agar, charged with lead nitrate (0.005 M). Potassium chromate (2, 1.5, 1 M) supernatant. After Kanniah *et al.* (1984). Contributed by P. Ramasamy. The ring spacing mostly decreases here, but at some stage near the bottom of the tube this trend is reversed, and the last few ring spacings actually increase.

would have to be carefully specified and controlled, to an extent to which, judging from the record, they never have been in the past.

5.4   Analytic formulations
Supersaturation models

Some of the most important endeavors to deal with the Liesegang Ring phenomenon algebraically have been made in the context of silver

Fig. 5.3.4. Closely spaced Liesegang Ring systems of copper chromate. (*a*) Alumina gel, Rings (see arrow) light. (*b*) Silica gel. Rings (see arrow) dark. Photomicrographs under partially polarized light. Contributed by Adair (1986).

(*a*)

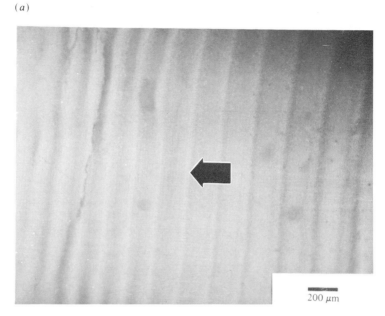

200 μm

(*b*)

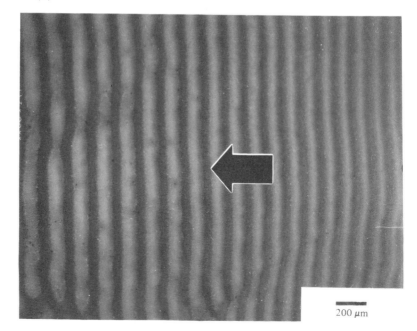

200 μm

chromate precipitation. Those are based on the supersaturation model, and Morse and Pierce (1903) explained that the same approach could be used equally well on many other examples, of which they mention, *inter alia*, $PbSO_4$, $Ag_4P_2O_7$, $AgCNS$, $AgBr$, $Co(OH)_2$, $BaCrO_4$ and $HgBr$.

Liesegang Rings of silver chromate owe their formation to the reaction

$$2AgNO_3 + K_2C_2O_4 \rightleftarrows Ag_2CrO_4 + 2KNO_3 \tag{5.4.1}$$

When the compound is present as a solid, the remaining undissociated chromate in solution must be in equilibrium with it. Hence,

$$[Ag^+]^2[CrO_4^{2-}] = K_s \tag{5.4.2}$$

where K_s is the solubility product, i.e. the concentration product which corresponds to saturation. However, we have already noted that precipitation does not automatically take place when K_s is exceeded, and Morse and Pierce knew this well enough. They did, however, link their treatment crucially to the notion that precipitation does take place automatically and immediately when the concentration product reaches a higher value, namely K_s'. The matter has already been discussed in Section 4.1, where it was seen that this is manifestly not a sufficient condition. In view of this, it is tempting to discount all the quantitative work of Morse and Pierce, but that would be too drastic a move; the approach still has some points of interest, as will be shown below.

The Morse and Pierce analysis refers to a particular geometry, as shown in Fig. 5.4.1(a). The vertical tube contains gel and dilute potassium chromate, concentration B_G. At a time $T = 0$, it is dipped into a strong solution of silver nitrate. The variable y represents the measured distance along the tube, A_R denotes a reservoir concentration, and A and B are, respectively, the concentrations of silver and chromate ions at time T. At $T = 0$, we have $A = A_R$ for $y < 0$, and $A = 0$ for $y > 0$. Correspondingly, at the starting time, we have $B = 0$ for $y < 0$ and $B = B_G$ for $y > 0$. The dissociation of potassium chromate is assumed to be complete, which is reasonable, on the assumption that B_G is small. The silver chromate is also taken to be completely dissociated. Two diffusion constants must be defined, D_A for silver ions and D_B for chromate ions. The cross-sectional area of the tube will be denoted by $\mathscr{A}$.

The amount of silver that passes a given boundary in time dT is given by

$$\mathscr{M}_A = -D_A \frac{\partial A}{\partial y} dT \mathscr{A} \tag{5.4.3}$$

at y, and the amount which reaches a boundary at $y + dy$ is

$$\mathscr{M}_A + \frac{\partial \mathscr{M}_A}{\partial y} dy = -D_A \frac{\partial A}{\partial y} dT \mathscr{A} - D_A \frac{\partial^2 A}{\partial y^2} dT \mathscr{A} \, dy \tag{5.4.4}$$

The amount of silver which accumulates between y and $y + dy$ is the difference, i.e.

$$D_A \frac{\partial^2 A}{\partial y^2} \, dT \, dy \mathscr{A} = dA \, dy \mathscr{A} \tag{5.4.5}$$

from which:

$$\frac{dA}{dT} = D_A \frac{\partial^2 A}{\partial y^2} = \frac{\partial A}{\partial T} \tag{5.4.6}$$

This expression is actually well-known as Fick's Law, already used in Section 2.3. Indeed, we could have begun here, but it is interesting to see how the relationship comes about.

In exactly the same way, we also have

$$\frac{dB}{dT} = D_B \frac{\partial^2 B}{\partial y^2} = \frac{\partial B}{\partial y} \tag{5.4.7}$$

Of course, it had to be assumed, contrary to the caution expressed above (Section 5.3), that the appearance of solid precipitates leaves the diffusion constants unchanged; indeed, only under these conditions can analytic solutions of any kind be obtained.

Fig. 5.4.1. Growth systems envisaged by analytic models. (*a*) Morse and Pierce (1903); compound (A_2B) (*b*) Wagner (1950); compound (AB)

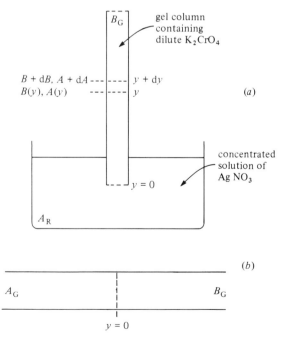

Though the last two equations are alike (indeed, they are the same), their solutions are different, because they are subject to different boundary conditions. Thus, for eqn (5.4.6) we have $A = A_R$ at $y = 0$ for all values of T (assuming an infinite reservoir of silver nitrate), and $A = 0$ for $y > 0$ at $T = 0$. For eqn (5.4.7) we have $B = B_g$ at $y > 0$ and $B = 0$ for $y < 0$ at $T = 0$.

The solutions which can be shown to satisfy these conditions:

$$A(y, T) = \frac{2A_R}{\pi^{1/2}} \int_{y/2(D_A T)^{1/2}}^{\infty} \exp(-z^2)\, dz \qquad (5.4.8)$$

and

$$B(y, T) = \frac{2B}{\pi^{1/2}} \int_{-y/2(D_B T)^{1/2}}^{\infty} \exp(-z^2)\, dz \qquad (5.4.9)$$

where z is merely an integration variable. A brief test of the solutions will suffice. Thus, if we take $T = 0$ in eqn (5.4.8), then the two integration limits are the same and the integral is therefore zero. If, at any other time, we put $y = 0$, then

$$A = \frac{2A_R}{\pi^{1/2}} \frac{\pi^{1/2}}{2} = A_R \qquad (5.4.10)$$

which is the boundary (driving) concentration for silver. Eqn (5.4.9) reflects the boundary conditions for chromate ions in a similar way.

If there is, indeed, a quantity K'_s as envisaged above, then it must, for silver chromate, satisfy.

$$K'_s = A^2 B = \frac{4A_R^2 B}{\pi^{3/2}} \left[\int_{y/2(D_A T)^{1/2}}^{\infty} \exp(-z^2)\, dz \right]^2$$

$$\times \left[\int_{-y/2(D_B T)^{1/2}}^{\infty} \exp(-z^2)\, dz \right] \qquad (5.4.11)$$

and, as Morse and Pierce pointed out, the task is to ascertain whether a K'_s so calculated is indeed constant for different values of A_R and B_G.

Because the values of D_A and D_B were not actually known to Morse and Pierce, these researchers first used an indirect method of testing the validity of eqn (5.4.11). They argued that a constant K'_s should imply constant (though not, of course, equal) values of $Y/2(D_A T)^{1/2}$ and $Y/2(D_B T)^{1/2}$, where Y is now the position of a ring and T the time of its first appearance. Morse and Pierce examined this relationship for several ring systems by measuring pairs of corresponding T and Y values, and a typical run is shown in Table 5.4.1. The numbers in the third column are indeed remarkably constant, and it almost looks as if the model were thereby vindicated in its essentials, but there are good grounds for scepticism on this point. Thus, it is clear that *some* material must have been withdrawn from the diffusion system by each ring as it formed, which means that the conditions envisaged by eqn (5.4.11) could have prevailed only for the first ring. That ring, as it happens, is often inaccessible to observation, being too near to the gel–liquid interface, but in the Morse and Pierce experiments, B_G was purposely kept low in order to shift the first ring to sensibly high values of Y. Morse and Pierce do not refer to the problem of concentration contour distortion by ring formation; the observed constancy of $Y/T^{1/2}$ is in a sense more of a mystery than it is a confirmation of anything in particular. One could argue that little material would be withdrawn by precipitation in a system for which K_s' exceeds K_s only by a small amount, but that very condition would make ring formation unlikely; see Section 5.6.

In any event, as Morse and Pierce realized, the constancy of $Y/T^{1/2}$ could not by itself prove the constancy of $A^2 B$ for precipitation. It could prove only that there is *some* function of $Y/T^{1/2}$ which is the right-hand side of eqn (5.4.11), without demonstrating that the left-hand side is A^2B. Morse and Pierce therefore attempted to take the exploration further (albeit under the same assumptions) by *assuming* K_s' to be constant, at least for very small variations of A_R and B_G and then solving pairs of

Table 5.4.1. *Typical observation by Morse and Pierce (1903). Growth of Ag_2CrO_4 in gelatin. $AgNO_3$ solution* $1N$; K_2CrO_4 *solution: $1/75$ N temperature:* $15.7\,°C$.

time, s	distance Y, cm	$Y/T^{1/2}$, cm s$^{1/2}$
1245	0.537	0.01522
1420	0.575	0.01526
1607	0.611	0.01524
1825	0.649	0.01519
2068	0.694	0.01526
2345	0.738	0.01524
2658	0.785	0.01523
3000	0.834	0.01523
3395	0.888	0.01524
3823	0.940	0.01520
4305	0.998	0.01521
4842	1.056	0.01518
5443	1.125	0.01524
6102	1.188	0.01520
6870	1.260	0.01520
Mean		0.01523

eqns (5.4.11) simultaneously. Advantage was taken of the fact that in the silver chromate system under review the diffusion of chromate ions was not rate-limiting, which made the exact value of D_B unimportant. That enabled D_A to be evaluated as $D_A = 1.56 \text{ cm}^2/\text{day}$. Again, both D_A (referring to silver) and K'_s did indeed turn out to be remarkably (if not totally) constant. When Morse and Pierce did note departures from their expectations, they ascribed them not to any basic shortcomings of the model, but to non-homogeneity of the (agar) gels they were using, and even to the non-constancy of the growth tube bore. The Morse and Pierce formulation proved its worth during these early experiments, not so much because it was correct, but because it seemed highly quantitative and therefore encouraging. Even so, no similarly extensive set of experiments appears to have been made.

By now there are several additional reasons for questioning the validity of the conclusions. Thus, in this and similar work of this kind, the observed times T at which rings are first seen, have been widely interpreted as symptoms of a certain critical supersaturation reached. This assumes that there is no 'overshoot' at all, i.e. that this degree of supersaturation is a sufficient as well as necessary condition, and that ring formation follows immediately, irrespective of the A/B ratio. More recent work has shown that this ratio actually has an important bearing on the time and position of ring formation. Thus, for instance, Meal and Meeks (1968) have noted that in their systems (barium sulfate growth in agar) precipitation was significantly delayed. The matter will be further discussed in Sections 5.5 and 5.6.

Researchers who followed Morse and Pierce (1903) made a variety of contributions, but these did not alter the general picture. Almost half a century was to elapse before the next major onslaught on the problem, by Wagner (1950). Wagner actually applied the same equations (he had no choice in that matter) but he was fully aware of the fact that the reagent concentrations would have to be sharply decremented at the precipitation point, by the precipitation itself. He argued also that at that point, given by $y = Y$, subsequent deposit growth would call for the counter-diffusion rates to be equal. Thus

$$-D\frac{\partial A}{\partial y} = D\frac{\partial B}{\partial y} \tag{5.4.12}$$

No distinction was made between the two diffusion constants D_A and D_B, both being taken as D, but in later work along similar lines, Gerrard *et al.* (1962) did introduce separate variables. Wagner accepted the notion of a critical supersaturation product K'_s as a necessary and sufficient precipitation condition (see, however, Section 5.5) and, indeed, had to make other assumptions in order to make the situation algebraically tractable. Thus, it was assumed that $A = B$ at $y = Y$ immediately before precipitation, which meant $A(Y) = B(Y) = 0$ (almost) immediately after precipitation. Wagner also postulated that once a deposit was formed and was steadily growing (undisturbed by any other deposits) one could take $A(y > Y) = 0$ and $B(y < Y) = 0$. The growth system envisaged by Wagner differed in minor ways from that shown in Fig. 5.4.1(a); Wagner considered a simple tube, suggesting that it might be filled with gel entirely, initially containing reagent (A) to a concentration A_G for $y < 0$, and reagent (B) to a concentration B_G for $y > 0$; Fig. 5.4.1(b). Under such conditions, the diffusion equations have solutions which differ somewhat from those given above, namely

$$A = \tfrac{1}{2}(A_G - B_G) - \tfrac{1}{2}(A_G + B_G)\ \mathrm{erf}[y/2(DT)^{1/2}]_{y < Y} \qquad (5.4.13)$$

and

$$B = -\tfrac{1}{2}(A_G - B_G) + \tfrac{1}{2}(A_G + B_G)\ \mathrm{erf}[y/2(DT)^{1/2}]_{y > Y} \qquad (5.4.14)$$

Setting $A = 0$ or $B = 0$ at $y = Y$, one obtains

$$\frac{1 - \exp(\gamma)}{1 + \exp(\gamma)} = \frac{A_G}{B_G} \qquad (5.4.15)$$

where $\gamma = Y/2(DT)^{1/2}$. For this γ, and thus Y, could be calculated as a

Fig. 5.4.2. Calculations by Wagner (1950) giving the position Y of the first precipitate as a function of the initial reagent concentrations A_G and B_G. System as in Fig. 5.4.1(b).

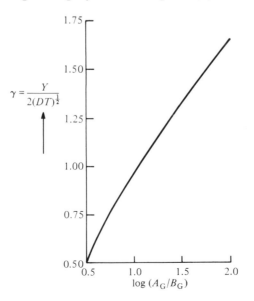

function of A_G/B_G, and Fig. 5.4.2 shows Wagner's results. Once again, these are less useful than one might have hoped, just because they refer only to the first ring. However, Wagner also came to qualitative conclusions concerning concentration profiles which are remarkably similar to the modern computer product. The computer solutions of Section 5.6 do, however, give a better insight into the matter, *inter alia* because they can make allowance for variable growth rates.

Wagner drew attention to the fact that if precipitation were to proceed at the same rate at which the supersaturation front advances, crystals in gels would grow as long needles, parallel to the *y*-axis. They do not do so and, indeed, the direction of diffusion has remarkably little influence on the direction in which individual crystallites grow. One concludes once again, as in Section 3.1, that these crystallites are surrounded by a mantle of uniform or almost uniform concentration, and that the growth itself is limited by the rate of surface nucleation. As a result, the advance of the supersaturation front is much faster than the crystal growth, and because the precipitation region acts as a 'sink' for solute at $y = Y$, no new precipitation can take place until the supersaturation front has advanced well beyond that region, to a place where the precipitation conditions are once again fulfilled.

Wagner actually calculated ring spacings on the basis of the above equations (and with a good deal of mathematical agony) but in view of the assumptions made, the conclusions have only schematic validity. Wagner recognized this, and mentioned also that the constant $\Delta Y/Y$ ratio cannot in fact be numerically evaluated, because the various parameters which enter into it are not known with a sufficient degree of reliability. Nor do we know more than a little about the way in which these parameters are influenced by the nature of the medium. How important this is can be seen from an observation by Veil (1935), who found that certain silver and cuprous salts form rings in gelatin, but not in agar, whereas lead iodide formed them in agar and not in gelatin! The effect of trace impurities is similarly vexing; see Foster (1919), and Schleussner (1922). The same could be said for the effect of pH; see Bradford (1916). One very familiar observation is reflected by Wagner's equations: other things being equal, dilute solutions make for widely spaced rings, but one would have guessed that anyway; see also Prager (1956).

Competitive particle growth models

It should be mentioned that supersaturation models in the form discussed above are certainly not the only ones to have been the subject of analytic work. For reasons of economy, one looks in the first instance for a single potent model, but we must ultimately recognize that this search can be taken too far; the published literature makes reference to varied observations which do not appear to fit harmoniously into any single picture. Thus, the fact that Liesegang ring formation is occasionally observed in chemically homogeneous melts, in gel-free solutions (Morse and Pierce 1903, Morse 1930), and also in bacterial cultures(!), suggests that quite different mechanisms can lead to outwardly similar results. Among vexing examples are findings by Ghosh (1930), according to which rings can form even when the system is precharged with dispersed crystallites. In their presence, supersaturation should not occur, and other mechanisms which might produce periodic precipitation would have to be sought. On the other hand, there remains the possibility that crystallites (with ill-characterized and possibly 'passivated' surfaces) might *not* prevent a degree of supersaturation. The prevention would have to be brought about by crystal growth, and that, in turn, would call for surface nucleation, which may or may not be immediate.

Interest in alternative approaches has nevertheless remained strong. Thus, for instance, Flicker and Ross (1974) pointed out that, because of surface tension, large crystallites are less soluble than small ones, and will therefore grow at the expense of the latter. Indeed, this should be true not only for crystallites, but also for any colloidal particles that might exist before crystallites as such are formed; see Marqusee and Ross (1984) and also Ortoleva (1978) and Feinn *et al.* (1978). Ortoleva (1984) applied these considerations to an idealized geochemical problem, and found likewise that 'competitive particle growth' can lead to the formation ('self-organization') of density striations. The descriptive equations had to be solved numerically, and some assumptions had to be made to establish the 'symmetry breaking' conditions which brought periodic solutions about.

A more detailed treatment has been given by Lovett *et al.* (1978), who obtained explicit solutions on the basis of their own pattern of assumptions. This approach is based on considerations developed in the course of earlier work, originally by Lifshitz and Slezov (1961), Wagner (1961), and later by Kahlweit (1975); see also Dunning (1973). Kahlweit's paper is particularly detailed and closely argued. These ideas concerned themselves with the diffusion-controlled growth rate of a spherical particle of radius r, that rate being given by

$$\frac{dr}{dT} = \frac{DC_\infty}{\rho r}\left(\phi - 1 - \frac{2\Sigma}{\rho r k t}\right) \tag{5.4.16}$$

where D is, in this case, the diffusion constant of the reaction product (or of the monomer, if no reaction is involved) and ρ the density of the precipitated particle. For simplicity, D has been taken as a simple constant. As always here, T is time, t the temperature, ϕ is the supersaturation, and Σ the surface free energy of the crystals. C_∞ is the true equilibrium concentration of the reaction product (or monomer), i.e. the concentration that would be in equilibrium with a particle of infinite radius.

It will be seen that dr/dT is positive only when $r > r_c$ such that

$$r_c = \frac{2\Sigma}{\rho\phi kt} \tag{5.4.17}$$

which is schematically plotted in Fig. 5.4.3. Only then can particles grow; otherwise they dissolve.

Lovett and co-workers applied this concept to an assembly of particles that is subject to a time-dependent particle size distribution $F(r, T)$ which, in the steady state, must satisfy

$$\frac{\partial F}{\partial T} + \frac{\partial F}{\partial r}\frac{dr}{dT} = 0 \tag{5.4.18}$$

Then there has to be a continuity relationship, such that

$$\phi(T) = \frac{4\pi\rho}{3C_\infty} \int_0^\infty Fr^3\, dr = \phi(0) \tag{5.4.19}$$

For a given initial distribution $F(r, 0)$ and a given starting supersaturation $\phi(0)$, the last two equations determine $\phi(T)$ and $F(r, T)$ at all times.

Fig. 5.4.3. Supersaturation necessary for growth, as a function of particle size; schematically after Lovett *et al.* (1978).

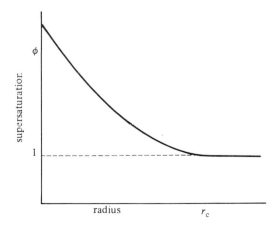

Lovett *et al.* proceeded to solve these equations for specific cases of *F*, after providing a qualitative assessment of the implications.

These implications are simple enough in principle. Fig. 5.4.4(*a*) shows the uniform solute distribution, in unstable (dynamic) equilibrium with crystallites. If there were a spontaneous fluctuation at some place, a few extra crystals might dissolve, releasing material which (in view of Fig. 5.4.3) would allow a few of the larger crystals to grow. As a result of that growth, the local supersaturation would diminish (Fig. 5.4.4(*b*)) and, as a secondary consequence, particles would diffuse towards that region from neighbouring regions. This would tend to raise ϕ above the value prevailing locally within the disturbance and would thereby further encourage crystal growth in that region. The prevailing tendencies are thus self-reinforcing, and lead to the situation shown in Fig. 5.4.4(*c*), which implies a spacing between deposits. However, the model is one-dimensional, and one must expect the conditions to randomize in directions parallel to the structure here envisaged. Of course, once an overall gradient is superimposed (as it is in normal gel systems), systematic

Fig. 5.4.4. Spontaneous development of structure by competitive particle growth in a one-dimensional system; $\bar{r}$ = local value of mean crystal radius; schematically after Lovett *et al.* (1978). (*a*)–(*c*) = increasing time.

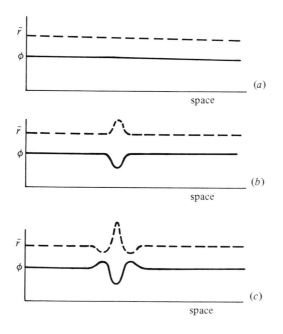

structuring is much more easily envisaged, though the situation is then deprived of some of its philosophical poignancy. In the absence of such a gradient, banding would be expected to occur (on the present grounds) only in *very thin* systems, i.e. those of a thickness small enough to be comparable with the longitudinal periodicity.

The model does not by any means make supersaturation irrelevant, but it relegates it to a supporting role. Its function is to permit the reaction product (or monomer) to nucleate in the first place, this nucleation being (by implication) simple and fast. Ring formation would then follow on other grounds. That competitive particle growth actually happens is an established fact; one can see it in crystal growth systems on a macroscopic scale (confirmed by time-lapse cinematography), but whether it is in any particular case due to the differential solubility of large and small grains or to the prevailing concentration contour pattern is another question. The fact that the crystals in each layer are more or less of the same size *is* in harmony with the model.

In general terms, and entirely outside the Liesegang context, the study of systems in which spatial inhomogeneities develop spontaneously under initially homogeneous conditions has a distinguished history. It was, for instance, a subject of investigation by Turing (1952) of 'Turing Machine' fame, also by Prigogine and Nicolis (1967), Glansdorff and Prigogine (1971) and, with special reference to diffusion, Fife (1976). As Flicker and Ross (1974) put it: 'for a system far from equilibrium the coupling of auto- or cross-catalytic chemical reactions with transport processes (diffusion) may bring about conditions in which the system becomes unstable, in that local (inhomogeneous) thermal fluctuations grow in time'. Moreover, similar consequences could also follow from non-thermal processes. The description of such a system (one, again, of the 'very thin' variety) by the authors was based on an initially homogeneous distribution of two ionized solutes. No macroscopic gradients were superimposed; indeed, the findings suggested that the presence of such gradients would not materially change the predicted periodicities. These predictions were derived analytically, albeit again only for a one-dimensional system, on the basis of rate equations which had to be linearized to make them solvable, but which nevertheless included terms for the adsorption and desorption of material on colloid particles, and also terms for the diffusion of the colloid particles which represent the preprecipitate form. The authors checked their predictions against experimental results, growing PbI_2 in agar, and using radioactive tracer techniques to measure concentrations. Some of their conclusions have already been discussed in Section 5.2.

Sol coagulation models

The notion that sol formation precedes ring formation has already been mentioned in Section 4.1, and a great deal of model-making along those lines can be found in the literature, much of it linked specifically to AgI, e.g. see Roy (1931), who used starch and silica as media, and Chatterji and Rastogi (1951), who used agar. How a sol particle is formed in the first place is not part of these models, and we may assume that some form of homogeneous nucleation is involved, just as it often is for crystallites. Similar energy considerations ought to apply.

We shall here assume that $AgNO_3$ has diffused into a KI gel for some time, and that sol particles have come to exist over a range of the medium without (at that stage) coagulating ('flocculating') anywhere, having been created by an advancing 'sol front' (Adair, 1920, Shinohara, 1970). The presence of such matter has already been referred to in connection with Fig. 5.2.6. Once formed, the particles will adsorb ions, in this case silver and nitrate ions, but not in equal numbers; Beekley and Taylor (1925) have shown that the nitrate ion is adsorbed less than the silver ion. As a result, the particles become positively charged, and this has been demonstrated by observations of cataphoresis (Dhar and Chatterji, 1925a). Accordingly, corresponding negative charges must surround the particles in the electrolyte medium. Those negative charges form Debye screening layers; their average charge density diminishes with distance. It is an *average* charge density, because there are fluctuations from the local equilibrium value, due to Brownian motion. The diffuse Debye layer ordinarily screens each particle from its neighbor, but if, in the course of Brownian movement, two sol particles were nevertheless to approach one another, their positive core charges would prevent them from actually colliding and thus from forming larger particles. Indeed, the surface charge is an essential condition of sol stability, and the contact prevention is effective, as long as the surface charge is high enough. Vincent *et al.* (1971) have studied the influence of silver adsorption on the stability of AgI sols, and have shown that stability diminishes with increasing distance form the gel boundary, see also Williams and Ottewill (1971). The adsorbed charge is lowered when a charged particle finds itself in surroundings in which the nitrate ion concentration is greater than the silver ion concentration. This actually happens in due course, as $AgNO_3$ continues to diffuse into the gel, because free silver ions are removed from the electrolyte by their preferential adsorption. When nitrate ions have reached a critical surplus concentration (sometimes called the 'flocculation value'), the sol particles are able to coagulate; a precipitate will form. Smoluchowski (1916, 1917) first proposed such a mechanism.

What happens immediately after the precipitation is still a matter of conjecture. According to some models, the precipitate itself adsorbs silver ions; according to others (e.g. Dhar and Chatterji, 1925a, b) it adsorbs nitrate ions. There is a similar uncertainty about the way surviving sol particles in the vicinity react to the fresh precipitate. Dhar and Chatterji believed that these particles come themselves to be adsorbed onto the precipitate and incorporated into it, being thereby removed from the neighboring gel region. Shinohara (1970) entertained the same idea, but believed that the sol particles as such are first dissolved under the influence of ions released in the flocculation zone. All this leaves a great deal of room for speculation. However, in one way or another, solute is removed from a gel region, and a new flocculation cannot take place until the whole process is repeated lower down the diffusion tube.

Without entering into the problem of the role played by the precipitate, Kanniah *et al.* (1981a) have used these notions to propose a model for 'revert' rings, e.g. as described by eqn (5.3.3). Because the sol particles are less and less positively charged at greater depths, the isoelectronic nitrate concentration which is necessary for flocculation is more and more easily attained. Thus, the spacing between rings ought to decrease. Further down still, the particles should actually become negatively charged due to excess nitrate adsorption, and the relationships should then be reversed. See Palaniandavar *et al.* (1985) for a discussion of charge reversal, but the matter remains to be investigated in greater detail.

Selective adsorption on which all these models depend is actually very common. Thus, for instance, Lee and Meeks (1971) have reported that Liesegang Rings of lead chromate exhibit membrane-like characteristics by selectively accumulating potassium and calcium ions, while allowing sodium, magnesium, and 'a wide variety of other inorganic ions' to diffuse freely. In much earlier work by Bradford (1926) it was demonstrated that even fully formed ring precipitates can still act as adsorption surfaces for colored additives, and in this way many barely visible deposits can be made distinct. Many other examples have been given by Dhar and Chatterji (1925a).

Models of Liesegang Ring formation which envisage an important role for sol flocculation may indeed by applicable in certain cases and, in a sense, they offer insights which go in potentially important ways beyond those offered by simple supersaturation theories. However, they are evidently complex, involving reagent diffusion, sol formation, sol particle diffusion, and adsorption, as well as electrostatic space charges, and a comprehensive analysis of the interplay between these components has not yet been attempted. However, separate parts of the process have been extensively studied on a quantitative basis, mostly in the context of colloid and flocculation research, e.g. see Kruyt (1952) and Everett (1973), and the details are beyond the scope of this book. Nevertheless, it will immediately be plausible that a variety of ring spacing 'laws' can be predicted by adopting different patterns of assumptions concerning adsorption, desorption and the dynamics of these processes, e.g. see Shinohara (1970, 1974), and Mathur (1961).

5.5   Precipitation conditions under supersaturation

The critical nucleus concept described in Section 4.1 tells us something about the intrinsic stability of agglomerates, but does not concern itself with the problem of how such agglomerates come about. They *can* come about under any conditions, but the conventional wisdom is that only when the concentration product exceeds the solubility product K_s are some nuclei of critical size and above formed. We shall take it here that we are dealing with a simple ionic compound of two components (A) and (B), in which case K_s simply equals AB. (For a different case, see Section 5.4). K_s actually has its origin in the Law of Mass Action. That is a good pedigree, but we know even then that crystallization is not automatic and inevitable even when $AB = K_s$. Supersaturation is a common experience, and only when the concentration product reaches a higher value K_s' is precipitation *from solution* actually spontaneous. Whether it is also *immediate* (i.e. at the moment of K_s' being reached) is another question, and one without a clear-cut answer. Two problems arise: what is the fundamental significance of K_s', and is $AB = K_s'$ by itself a *sufficient* condition of precipitation in a stochastic system like a gel.

The gel literature abounds with papers which are based on the assumption that $A(X, T)B(X, T) = K_s'$ marks the position X along a growth tube at which the first Liesegang Ring is formed, e.g. see Morse and Pierce (1903), Wagner (1950), Meeks and Veguilla (1961), Gerrard *et al.* (1962), Venzl and Ross (1982), and Ortoleva (1984). In all these cases, the supporting arguments relied on analogies with precipitation from solution, i.e. systems of uniform concentrations. If this criterion were correct then, in a system of the kind shown in Figs. 2.3.1 and 5.6.2, the first precipitation (i.e. the first Liesegang Ring) would always have to take place where the concentration product is a maximum. It has already been shown (see Section 2.3) that this maximum always occurs in the middle of the system ($X = L/2$), no matter what the boundry concentrations of (A) and (B) might be. Many observers (e.g. Kirov, 1969, 1972, 1980) have found otherwise, but have either been unaware of the implications, or else have regarded them as a direct consequence (only) of a difference between diffusion constants. The matter was first brought to wider attention by García-Ruiz and Miguez (1982), who pointed out that the position (X) of the first precipitate is in fact dependent on the reservoir concentrations A_R and B_R. This important fact had escaped notice, precisely because the simple gel-growth systems in wide use did not lend themselves to highly critical tests.

By now it is no longer plausible to believe in $AB = K'_s$ as a sufficient condition for precipitation, although it can be shown to be a necessary one. This demonstration can be provided in a number of ways. Thus, modern theories that concern themselves with the dynamics of particle agglomeration, e.g. Zettlemoyer (1969), suggest that the probability P_N of nucleus formation is dependent on the supersaturation ϕ in accordance with an expression of the form:

$$P_N = \exp\left[-\frac{b}{(\log \varphi)^2} \right] \tag{5.5.1}$$

where b is a constant depending on the morphology of the building blocks and on the surface energy of the clusters. The term $\log \phi$ may be considered the driving force of the nucleation process. The supersaturation itself is defined as $\phi = AB/(AB)_s = AB/K_s$, where $(AB)_s$ is the concentration product at saturation. Eqn (5.5.1) has the well-known form shown in Fig. 5.5.1(a). Though P_N must eventually saturate (since it cannot be greater than unity), it is clear that the probability of nucleus formation is negligible up to a certain value of $\phi = \phi_c$. At that point there is a virtually abrupt start, and it can therefore serve to define a quantity K'_s, such that $\phi_c = K'_s/K_s$. Precipitation is possible at lower values of ϕ, but is then highly unlikely. This constitutes the link between schematic, semiempirical notions of K_s, and modern microscopic models; see also Henisch and García-Ruiz (1986b). Without being concerned about the origins of the process, Kurbatov (1931) confirmed the dependence of precipitation rates on the reagent concentration ratio. We do not, in general, have reliable information about K'_s for particular systems. For silver chromate, Van Hook (1938a) gives $K'_s/K_s = 5$, but it can be very much higher.

For precipitation from solution this approach appears to be entirely satisfactory, but despite the superficial successes scored by the K'_s hypothesis in connection with gels (e.g. see Section 5.4), and for the reasons given above, $\phi > \phi_c$ is evidently an insufficient condition for precipitation from gels. This condition suggests, for instance, that the A/B ratio is irrelevant, which is absurd for any system (e.g. one involving gels) limited by particle mobility. In this respect the situations which prevail in homogeneous mixtures, (e.g. as encountered in many laboratory and industrial processes involving solutions) and in gels are quite different. In a gel we are dealing with time-dependent macroscopic concentration gradients in which A/B takes on every possible value $0 < A/B < \infty$ along the gel column. This characteristic difference between static and stochastic systems has been widely overlooked in the formulation of Liesegang Ring models; a more realistic assessment of precipitation conditions demonstrates its crucial importance. Indeed, it has been pointed out (Venzl and Ross, 1982) that even computer solutions predict only continuous precipitation when the concentration ratio A/B is ignored.

That the A/B ratio does, in fact, matter is clear from the simplest and most schematic of models (Henisch and García-Ruiz, 1986b), relying on

Fig. 5.5.1. Conditions of precipitation: (*a*) Nucleation probability as a function of supersaturation; form of the relationship predicted by Zettlemoyer (1969). Here $b = 1$. (*b*) Schematic, two-dimensional (30×30) simulation of nucleus formation; probability of finding a

$$\begin{matrix} (B)(A) & & (A)(B) \\ & \text{or} & \\ (A)(B) & & (B)(A) \end{matrix}$$

configuration as a function of $A/(A+B)$. Results of three runs. After Henisch and García-Ruiz (1986b).

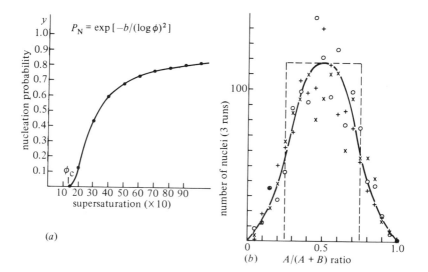

(*a*)

(*b*) $A/(A+B)$ ratio

geometry alone; all energetic and three-dimensional space related arguments would be additional. We may, for instance, populate a rectangular matrix of cells randomly with (A) and (B) items and ask about the probability of finding an

(A)(B)   (B)(A)
(B)(A)   (A)(B)

configuration that we could regard (here) as a 'nucleus'. That probability goes through a sharp maximum at $A = B$, and tends to zero rather rapidly as A/B departs significantly from unity. Fig. 5.5.1(b) shows the results of such a simulation. For practical purposes, the bell-shaped curve could, in the first instance, be approximated as a rectangle. Without taking the geometrical model too seriously, it is at once plausible that the precipitation probability in a gel should be very small, unless the reagent concentrations each have certain minimum values. How these minima arise is shown in Fig. 5.5.2, which also reflects the fact that small departures from $A = B$ are allowed ('equality range'), but large departures are forbidden; $AB = K_s$ represents the solubility curve; $AB = K_s'$ marks the boundary of likely precipitation. In a typical situation, K_s' might be exceeded, and some higher value $AB = K_s''$ would then mark the *range* of concentrations within which precipitation is likely to take place; see Fig. 5.5.2. The two asterisks denote the minimum and maximum A/B ratio, and as K_s'' tends to K_s', these boundary values must tend to coincide,

Fig. 5.5.2. Concentration relationship I. K_s = solubility product. K_s' = precipitation product. $K_s'' > K_s'$. Definition of the 'equality range', leading to A_{min} and B_{min}.

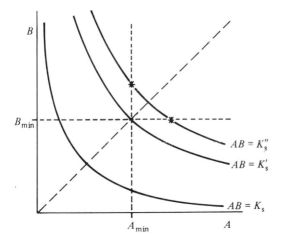

leading to $A = B = A_{min} = B_{min}$ as (for the simple case envisaged) the second precipitation criterion.

A three-dimensional simulation would show the equality range to be narrower than Fig. 5.4.1(b) suggests. Even then, only half the 'equality range' is actually effective. This will be clear from Fig. 5.5.3, in which the full curve represents the actual A and B value pairs between $x = 0$ and $x = L$, at a time when the first precipitation point is reached. Only half the range of values which satisfy the A/B ratio criterion is actually above the K_s'' curve. The outcome is that when a precipitation point is approached from the side of the lower boundary concentration A (as it is in the computations of Section 5.6) the second precipitation criterion in simple form is still correct, despite the symmetry of Fig. 5.5.1(b), and

Fig. 5.5.3. Concentration relationships II. 'Equality range' here assumed to be from $A/B = 0.8$ to $A/B = 1.2$. Full curve: actual relationship between solute concentrations A and B at the time (and place) when $A = B = 50$, for $A_R = 100$, $B_r = 200$. Broken curve: the concentration product AB at that time. After Henisch and García-Ruiz (1986b).

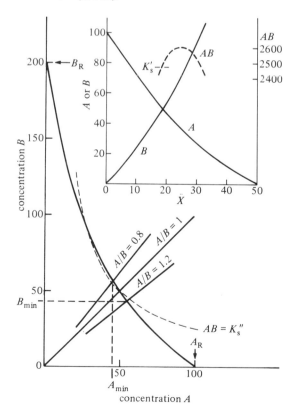

despite the unsharpness of B_{min}. If the concentration contours were more complicated, e.g. as they are expected to be after the first precipitation, this simplification would be impermissible. In all this we assume that rising concentrations in the course of time do not significantly overshoot the criterion for the first precipitation; constant equality range.

Fig. 5.5.3 further shows that when $AB = K_s''$ is only just satisfied, the equality condition has to be accurately fulfilled, but when $AB = K_s''$ precipitation is possible over a whole range of A/B values. Within that range, precipitation is, however, most likely at $A = B$. In the course of double diffusion (see Fig. 2.3.5), the actual concentration product increases with time everywhere, and can greatly exceed K_s' (see Fig. 5.5.3 insert), but unless there is within the gel a locality which also falls into the equality range, precipitation cannot take place. When it does, its consequences depend, of course, on the concentration product contours in the vicinity of the precipitation point. In earlier work, Kirov (1977) did note that the first precipitate always occurs within a certain distance *range* for different A_R and B_R values, without, however, addressing the concentration ratio issue or, indeed, the stochastic nature of the processes involved. In passing: Salvinien and Moreau (1960) and Lendvay (1964) ascribed a significant role to the ratio of diffusion coefficients (here taken as equal for simplicity).

5.6   Computational analysis
General considerations and hypothetical experiments

It will have to be admitted that a highly detailed and comprehensive computer analysis of Liesegang Ring phenomena is not yet available, but a useful beginning has been made, one within microcomputer scope. It is based on the diffusion algorithm discussed in Section 2.3, and on the precipitation criteria of Section 5.5; see Henisch and García-Ruiz (1986a, b) and Henisch (1986). The algorithm is applied to two simultaneously counter-diffusing species. A variety of nucleation conditions can then be imposed. The calculations envisage constant as well as depleting reagent reservoirs. Provision can easily be made not only for the growth of precipitates (at varying rates) after their initial formation, but also for their re-solution at any point at which the concentration product falls below K_s. Among the most important aspects of the system are the facts that precipitation leads to solute depletion (see also Keller and Rubinow, 1981), while solute depletion can lead to re-solution. Local 'solute depletion' is here meant to denote *total* depletion of the less concentrated component at the time of precipitation, but the continuation of diffusion processes after that time make the total depletion a purely temporary affair. In any event, the algorithm could easily be adjusted to make the depletion less than total. Re-solution, in turn, implies solute augmentation, and solute liberated in this way can later (after diffusion spreading) support new precipitation and growth elsewhere. All this happens while the main concentration fronts advance throughout the medium. The net outcome depends in any particular case on a delicate balance between these processes.

The nucleation algorithm can be taken as deterministic or probabilistic. The system is thus described by many parameters (numerically independent of one another as far as the computer program is concerned though not, of course, in reality) interacting in ways that are sometimes obvious, but more often subtle. All affect ring positions, spacings and thicknesses. A given ring may dissolve on one side and grow on the other; to the observer this will appear as ring 'movement', Indeed, Kirov (1972) has reported seeing 'oscillating movements'.

In the first instance, one might use the computer to describe a typical form of behavior, as shown in Fig. 5.6.1; note the use of dimensionless

Fig. 5.6.1. Computed concentration contours (a), and concentration products (b), in the vicinity of a precipitation point $X = N\Delta X$, here with $\Delta X = 1$. Typical case. Dimensionless variables. Full lines: immediately before precipitation at $X = 40$. Broken lines: soon afterwards. $A_R = 100$, $B_R = 200$. After Henisch and García-Ruiz (1986b).

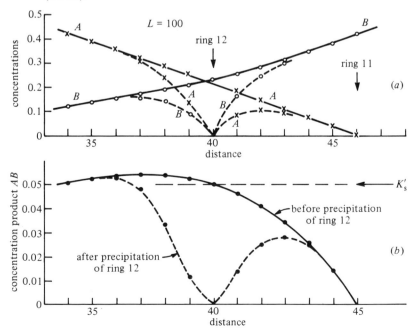

variables here and in all similar representations. Here, and indeed elsewhere in this section, the diffusion constants of the reacting species are
assumed to be equal (though, within limits, some allowance for other
conditions can be made). For a given system, Fig. 5.6.1 gives concentration contours, before and after the precipitation of ring number 12, ring
11 having already been precipitated. Ring 12 is due to appear at the
place ($X = 40$) where $A = B$ and $AB \geqslant K'_s$. As a result of its appearance,
the local concentrations are (here) totally depleted, and for a time soon
afterwards we see the consequences of the location. $X = 40$ acting as a
continued sink. In this way, and with varying degrees of sophistication,
concentration contours can be obtained for the whole system as a function
of time.

The more detailed operation of the available (one-dimensional)
algorithm is best described on the basis of hypothetical 'experiments',
in which parameters are varied one at a time. In this way, their action
can be explicitly demonstrated. To be sure, this is a privilege which
analytic solutions have always offered, but with those mostly ruled out,
the proper comparison here is with laboratory experiments.

The particular experiments here described are carried out in one or
other of the growth systems shown in Fig. 5.6.2. In some, the gel may
be pre-charged with reagent (A) to a uniform concentration A_G. In
others, we may have $A_G = 0$. Reagent (B), or reagents (A) and (B), diffuse
into the gel from reservoirs with initial concentrations A_R and B_R, whether
the gel is pre-charged or not.

Because the total number of adjustable parameters is so large, only a
few variations can be considered here, selected because they correspond
to widely familiar observations. The considerations are aimed at simple
principles, and no attempt will be made to reflect and describe experimental situations with specific reagents. As in Section 2.3, we take the
total length of the diffusion tube to be $L\Delta X$. It will again be convenient
to consider $\Delta X = 1$ (in dimensionless terms), making X and N (when
used) numerically equal. The limited resolving power of these computations should be borne in mind. Thus, the immediately adjacent masses
of successive (integral) values of X do not constitute separate rings; see
also Dee (1986).

Experiment 1: *Conditions of quasi-stability*
Fig. 5.6.3 shows a set of computed Liesegang Rings (based on the
configuration of Fig. 5.6.2(*a*) with $L = 50$) reached quasi-stability after
a long time ($T = 4000$). In the ordinary way, ring systems continue to
change with time at varying rates, but under the conditions depicted,

those rates have become very low. At large values of X, the concentration product is evidently very low, and some re-solution *is* recorded there, but the amount is small. Additional rings cannot form, because $AB < K_s'$ at small values of X, and the A/B ratio is far from unity. In this quasi-stable situation, growth and re-solution rates are practically balanced. However, it is also clear that they may not at this stage be *exactly* balanced. Rings at high X values must eventually begin to disappear. The solute released in this way will diffuse partly into the (B) reservoir, where it might be considered 'lost', and partly down the gel column, thereby increasing the AB product and changing the A/B ratio. No new nucleation is at all likely, but surviving rings could grow at the expense of the redissolving ones. Nevertheless, there would be a net loss of the

Fig. 5.6.2. Gel systems here used for hypothetical experiments; column length $= L\Delta X$. It is convenient to put $\Delta X = 1$. Dimensionless variables. (*a*) Closed at one end, gel initially charged with reagent (A) to a concentration A_G; liquid reservoir for (B) of concentration B_R, considered constant unless otherwise stated. (*b*) Double diffusion system; initial gel charge A_G may be zero; reagents (A) and (B) diffusing in from opposite sides, from reservoirs of initial concentrations A_R and B_R, regarded as constant unless otherwise stated.

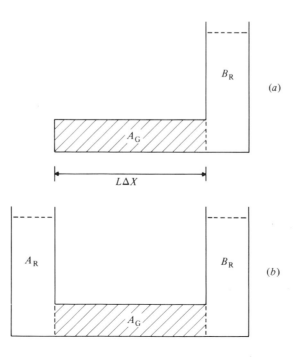

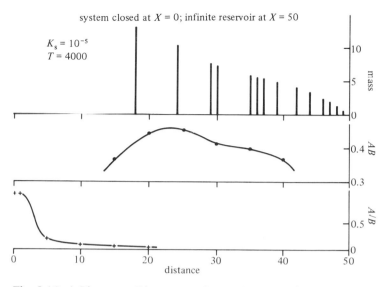

Fig. 5.6.3. A Liesegang Ring system in quasi-stability (Fig. 5.6.2(a)
configuration). $A_G = 1$, $B_R = 10$, $L = 50$, $K'_s = 0.5$. Deterministic
algorithm. Dimensionless variables. After Henisch (1986).

(A) reagent, and that loss would continue. The process may take a very
long time, but ultimately all the rings are expected to disappear if the
(B) reservoir is large; hence *quasi*-stability, rather than true permanence.

Experiment 2: *The role of precipitate solubility*
Figs. 5.6.4, and 5.6.5 show how ring systems (in a closed-tube configuration) develop for different solubility products K_s assuming an infinite
reservoir for (B). When the precipitate is only sparingly soluble, as in
Fig. 5.6.4, a highly structured ring system advances down the growth
tube. When the precipitate is much (e.g. ten times) more soluble, as in
Fig. 5.6.5, old precipitates (i.e. those formed early in the process) redissolve while later ones are still in the process of formation. The released
solute here supports the formation of new rings. Indeed, we have here
a pattern in which new deposits form not so much from primary reagents,
but at the expense of old deposits. Because old deposits do not entirely
disappear (within the times envisaged here) the entire tube gets filled to
an almost uniform precipitate density, and ring formation is suppressed.
When the precipitate solubility is intermediate, the results are also intermediate (Henisch 1986).
    The solubility product K_s actually affects the situation in two distinct
ways: (a) by allowing formed deposits to redissolve when they find

Fig. 5.6.4. Progressive development of a precipitation system for a solubility product $K_s = 10^{-5}$. Configuration of Fig. 5.6.2(a). $A_G = 1$, $B_R = 10$, $L = 100$, $K'_s = 0.5$, at different times, T. Distance $X = N\Delta X$, here with $\Delta X = 1$. Deterministic algorithm. Dimensionless variables. After Henisch (1986).

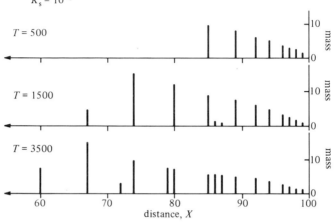

Fig. 5.6.5. Progressive development of a precipitation system for a solubility product $K_s = 10^{-4}$. Parameters as for Fig. 5.6.4. Dimensionless variables. After Henisch (1986).

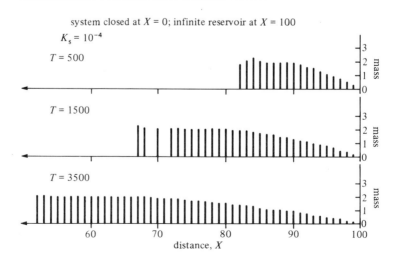

themselves in an environment in which $K_s > AB$, and (b) by exercising a controlling influence over the *growth rate*, here taken to be proportional to $AB/K_s - 1$. (The constant of proportionality may be called the growth factor). Either of these processes can be the dominating one in different situations. The growth factor is operative, as long as there is a sufficient amount of solute in the neighborhood. As this solute becomes exhausted, the growth rate comes to be controlled by the replenishment processes. Those are (i) the local formation of new (AB) material through the diffusion influx of (A) and (B), and (ii) the diffusion influx of material arising from the re-solution (partial or complete) of previously formed rings. The 'zero-sum' character of the situation is at the root of all observed changes. As noted above a ring may redissolve on one side and grow on the other; to the observer it will thus seem to move. The changes which occur, for instance, between Figs. 5.6.7(b) and 5.6.9 could be seen as a movement, and this kind of process is, in fact, frequently observed. Hermans (1947), for instance, reported on moving sulfide deposits.

The situations which occur in practice (as distinct from computer models!) are actually more complicated, in view of the well-known fact that large particles are less soluble than small ones; see Section 5.4. No single 'solubility', as here envisaged, can take care of these processes.

Experiment 3: *Effect of the growth rate*
When the growth rate is small, the solution is not rapidly depleted. A nucleation front can then advance through the medium, and the final effect will be that nucleation (followed by only slow growth) will happen over a wide band of the growth tube. In contrast, a high growth rate 'consolidates' each nucleation event as it happens, and creates large depletion regions in the neighborhood of any precipitate formed. As a result, and other things being equal, the precipitation should be more structured, and the computations confirm this (Fig. 5.6.6). The differences between the two forms of behavior are pronounced in the early stages, but become somewhat blurred later on, as more and more material diffuses into the gel and forms new deposits.

In the available algorithm, the growth rate is dependent on the supersaturation, and shows itself as a fractional increase of the existing mass in any location. In Kahlweit's (1960) (not fully transparent) analytic work, it is made dependent on the surface area of the existing deposit, taking each particle as spherical. This would be more appropriate, but it involves uncertainties of its own, especially since we know the accumulated mass, but not its state of distribution. On the basis of his approximations, Kahlweit (1962) concludes that ring formation is possible only

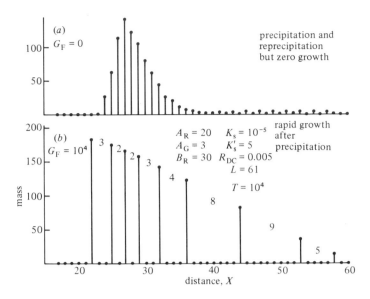

Fig. 5.6.6. Effect of the growth rate factor G_F on the structure of deposits, with reservoir depletion. Configuration of Fig. 5.6.2(a), with gel precharged to $A_G = 3$, $L = 61$, R_{DC} = reservoir depletion coefficient 0.005. Dimensionless variables. (a) Repeated precipitation allowed, but no subsequent growth of precipitates. (b) Repeated precipitation allowed, as well as high growth rate of precipitates already formed, subject to the 'zero-sum' requirement. Deterministic algorithm. After Henisch (1986).

when the growth rate is small, but it is hard to see why this should be so. Certainly, the finding is in direct conflict with the computer-derived results in Fig. 5.6.6.

Experiment 4: *Position of precipitate under double diffusion*
When A_R and B_R are equal (and the corresponding diffusion constants are likewise equal) then a growth system of the kind shown in Fig. 5.6.2(b) should yield a symmetrical deposit, whether continuous or in the form of Liesegang Rings. When $B_R > A_R$, the main deposit should be shifted to the (A) side. This prediction rests on the assumption that $A/B = 1$, where and when nucleation takes place. However, as discussed in Section 5.6, nucleation is *possible* for A/B values which differ appreciably from unity, albeit with (much) lower probability. This fact tends to shift the predicted ring towards the middle, where AB is a maximum, and where $AB > K_s$ is first satisfied. How important this shift turns out to be must depend on the growth rate factor, as discussed above, and on the rapidity with which the concentration front advances.

Given a totally deterministic algorithm, the first precipitate will always be predicted to occur in the middle, just because *some* nuclei will be formed there, no matter how far A/B is from unity. In Fig 5.6.7(a), the first deposit recorded was at $X = 31$, though, by the time $T = 600$ two small adjoining deposits have also formed. It is, however, important to draw a distinction between this early deposit and the 'main' precipitate that develops from it. The rudimentary layer of Fig. 5.6.7(a), which may or may not be visually detected in an actual growth tube, becomes in due course part of a larger complex, of which Fig. 5.6.7(b) shows a later stage, and Fig. 5.6.9(a) something close to completion. The peak distribution is then shifted distinctly to left-of-middle. This peak is likely to be perceived and recorded as *the* position of the Liesegang Ring, all earlier developments nothwithstanding.

Fig. 5.6.7. Stages in the formation of a quasi-stable deposit, with unequal (constant) reservoir concentrations. Deterministic algorithm Growth factor $= G_F = 1$. Dimensionless variables. (a) $T = 600$; first deposit formed (earlier) at $X = 31$. (b) $T = 1000$; distinct beginning of a 'center of gravity shift' to the left. See full contour in Fig. 5.6.9 for quasi-stable result. After Henisch (1986).

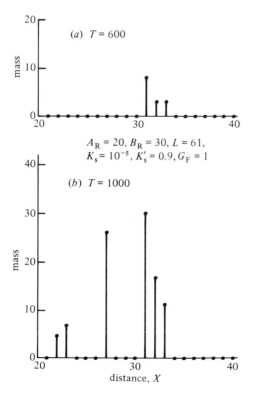

$A_R = 20, B_R = 30, L = 61,$
$K_s = 10^{-5}, K'_s = 0.9, G_F = 1$

The computer results thus lend support to conclusions reached by Ross (1984) on the basis of light-scattering experiments, namely that 'ring formation is a post-nucleation process'. The stage during which essentially no further nuclei are (locally) formed, but larger crystallites grow at the expense of smaller ones, is sometimes (but not with perfect fidelity) called 'ageing', e.g. see Lovett *et al.* (1978), and sometimes 'Ostwald ripening', e.g. see Kahlweit (1975). This stage occurs not only in Liesegang Ring development (copious nucleation), but undoubtedly also in systems in which only a few crystals nucleate.

When $A_R = B_R$, the deposit is, of course, symmetrical, as shown in Fig. 5.6.8. Distinct rings are formed during the early stages, but the vacant sites between them are filled in due course, as more material diffuses from the (here 'infinite') reservoirs. Once again, it can be shown that the precipitate solubility, expressed as $K_s = 10^{-5}$, plays a notable role in making the precipitate distribution continuous. With $K_s = 10^{-6}$, for instance, and otherwise identical conditions, the precipitate is structured.

Fig. 5.6.8. Stages in the formation of a quasi-stable deposit, with equal (constant) reservoir concentration. Deterministic algorithm. Growth factor $= G_F = 1$. Dimensionless variables (*a*) $T = 4000$; elements of structure present. (*b*) $T = 8000$; continuous (though sharply peaked) deposit. After Henisch (1986).

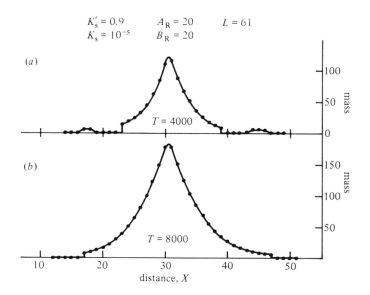

Experiment 5: *Effect of reservoir depletion*

It is a simple matter to make allowance for reservoir depletion and mutual contamination. For this purpose, the (B) reservoir, for instance, is *depleted* by $R_{DC}\,(dB/dX)_L$ during each unit of time, B (and thus dB/dX) being time-dependent. R_{DC} is a constant ('reservoir depletion coefficient'), dependent on the size of the reservoir. In addition, the (B) reservoir is depleted by $R_{DC}(-dA/dX)_L$, to take account of the influx of (A), which causes precipitation inside the reservoir, and thereby removes (B) material from it. Fig. 5.6.9(*b*) shows how drastic the corresponding changes of deposit structure can be. It is, moreover, abundantly clear that reservoir depletion is an important stabilizing factor in Liesegang Ring systems, inasmuch as it counteracts the tendency for new rings to be formed in the spaces between old ones.

Experiment 6: *Effect of unequal diffusion constants*

In all other experiments reported here, the diffusion constants D_A of species (A) and D_B of (B) are taken as equal, but the algorithm permits their ratio to be varied (within limits). As one would expect, these parameters have an important effect on the resulting mass distribution. Fig. 5.6.10 gives a typical example.

Fig. 5.6.9. Effect of reservoir depletion. Dimensionless variables. (*a*) $R_{DC} = 0$; zero depletion; prominent continuous peak. Middle of system. $X = 30.5$. (*b*) $R_{DC} = 0.005$; substantial depletion.

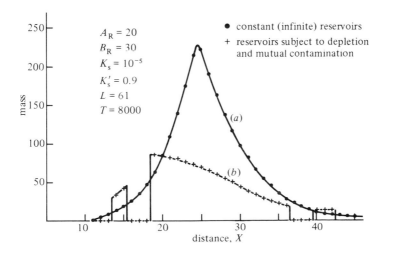

Fig. 5.6.10. Effect of unequal reagent diffusion constants (and thus mobilities): $D_A = 2D_B$. Deterministic algorithm; dimensionless variables. Configuration of Fig. 5.6.2(b). (*a*) High A_R and high D_A lead to deposition at higher values of X. (*b*) High B_R and high D_A lead to a skew version of the corresponding curve, Fig. 5.6.9(b).

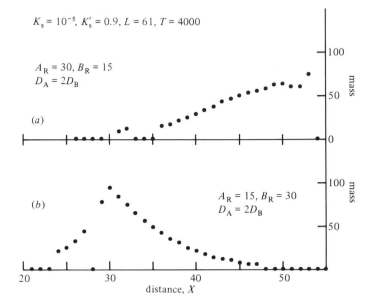

$K_s = 10^{-5}$, $K_s' = 0.9$, $L = 61$, $T = 4000$

$A_R = 30$, $B_R = 15$
$D_A = 2D_B$

(*a*)

(*b*)

$A_R = 15$, $B_R = 30$
$D_A = 2D_B$

distance, X

Fig. 5.6.11. Effect of the precipitation product $AB = K_s'$; comparison of precipitate distributions at $T = 2000$ for $K_s' = 0.5$ (full lines) and $K_s' = 5.0$ (mixed lines only), under otherwise identical conditions. Semideterministic algorithm; high growth rate, dimensionless variables. Configuration of Fig. 5.2.6(*a*). After Henisch (1986).

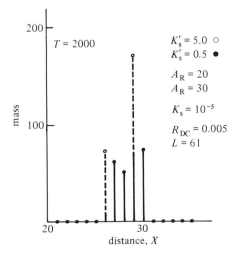

$T = 2000$

$K_s' = 5.0$ ○
$K_s' = 0.5$ ●

$A_R = 20$
$A_R = 30$

$K_s = 10^{-5}$

$R_{DC} = 0.005$
$L = 61$

distance, X

Experiment 7: *Effect of the precipitation product K'_s*
A high value of K'_s makes precipitation harder; specifically it tends to make precipitation more difficult within the depletion region near a previous precipitation, and thereby increases the ring spacing. Fig. 5.6.11 illustrates this for two values of K'_s, under otherwise identical conditions. For the lower K'_s a band is produced, for the higher value only two narrow rings. The total amount of mass precipitated is similar, but the distribution is very different.

Experiment 8: *Deterministic versus probabilistic algorithm*
All the experiments discussed above were based on deterministic algorithms: things happened immediately when the conditions were right, because they had to. It is, however, possible to build random elements of various kinds into the procedure. In practice, some such random element is evidently operative, and shows itself through the fact that nucleation has a built-in time-delay, presumably small when $(A-B)/(A+B)$ is small, and large when $(A-B)/(A+B)$ is large. Van Hook (1938a, b) has documented such delays, even in the presence of homogeneous (silver chromate) seeds. This kind of behavior is easily simulated. Fig. 5.6.12(*a*) gives a schematic representation of a ring, which we may regard as composed of separate deposits, one within each section ΔX, each the outcome (within limits) of chance. In that sense, the ring we see would be the sum-total of many deposits, as suggested by Fig. 5.6.12(*b*). The result is, of course, a smooth deposit contour, and this is demonstrated in Fig. 5.6.13. The results in Fig. 5.6.13(*c*) are based on a 'linear random element', but it is entirely possible to make the random aspect non-linear

Fig. 5.6.12. Schematic representation of probabilistic deposit formation. After Henisch (1986).

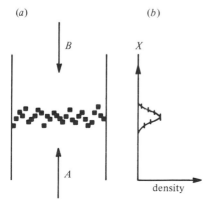

(Fig. 5.6.13(*a*)), so as to make the nucleation delay superlinearly greater, as small values are approached on the probability curve of Fig. 5.5.1(*b*). According to Kahlweit (1960), the number of crystals (of AgCl) nucleating diminishes when the reagents mix very quickly, and the probabilistic algorithms here envisaged take this fact implicitly into account. In any event, the smoothing out compared with the deterministic model is very obvious. Different ring positions arise once again from 'zero-sum' considerations, and the available algorithm makes, as yet, no provision for *lateral* diffusion. Kai *et al.* (1982, 1983) have reported that observed ring formations are in fact characterized by increasingly probabilistic spacings, as the reagent reservoir concentrations diminish.

Experiment 9: *Effect of precharging the gel*
In the experimental configuration of Fig. 5.6.2(*b*), the gel may be initially empty, or else it may be precharged to a concentration A_G, and Fig. 5.6.6

Fig. 5.6.13. Comparison between deterministic and probabilistic algorithms. $K'_s = 0.5$, $L = 100$, dimensionless variables. Configuration of Fig. 5.6.2(*a*). After Henisch (1986).

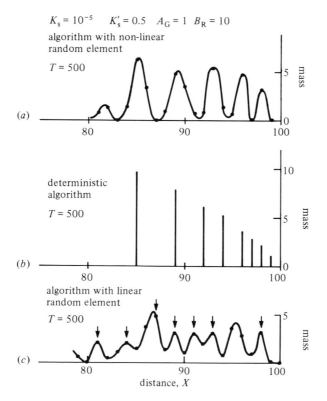

inter alia shows the corresponding results. In the ordinary way, precharging the gel makes the configuration of Fig. 5.6.2(b) look (initially) like that of Fig. 5.6.2(a) to the (B) reagent. Accordingly, ring formation begins on the right, and moves towards the left. However, the (A) reagent does not become exhausted, since it is replenished by its reservoir. As long as the supply of (A) is governed by the resident concentration A_G, the result is as usual a ring system with spacings increasing towards the left. However, at smaller values of X, that supply is governed by the (plentiful) (A), and there the ring spacings *diminish* towards the left. Such a reversal of spacing trends is occasionally observed in actual systems, and though these do not always conform to the geometries here discussed, at least one basic origin of such 'revert' patterns is now clear (see Kanniah *et al.* 1981a, b Kanniah 1983, Packter 1955, Flicker and Ross 1974, Mathur and Ghosh 1958). Indeed, it is possible for more than one such reversal of the spacing order to occur. There must, of course, be other mechanisms which produce such results. Kanniah *et al.* (1981a, b), for instance, have proposed one in terms of electrical charges on sol particles, and their effect on coagulation; see also Section 5.4.

Experiment 10: *Time relationships*
In Section 5.4 a certain (cautious) expectation was raised to the effect that rings at positions Y_N (or, as here, X_N) should appear at times T_N, such that $X_N / T_N^{1/2}$ is constant. Because of the approximations introduced by the analysis, this expectation was by no means exact, but it is nevertheless interesting to examine the extent to which the numerical computations are in agreement with this 'law'. The answer is that, as far as they go, they are in excellent agreement, indeed, better than one had any reason to expect. Fig. 5.6.14 shows this for two values of K_s'.

García-Ruiz, Santos and Alfaro (1987) have reported crystal masses which exhibit small periodic variations of growth rate while the diffusion is going on. New algorithms now available for the analysis of Liesegang Ring formation likewise demonstrate the existence of such 'growth waves', as shown in Fig. 5.6.15. The manner in which the size and frequency of the steps depends on a large number of control parameters, remains to be fully investigated. On the other hand, the principles are clear: the 'growth waves' arise, as do so many of the features discussed above, from the complex, zero-sum interaction between the primary diffusion processes on the one hand, and grain growth as well as grain re-solution on the other. The situation is also governed by the fact that events at one deposit site influence what happens at neighbouring sites. The results have some general implications for our understanding of diffusion controlled reactions. Periodicity as such is inherent in the diffusion–nucleation–growth process. It shows itself in terms of space as ring formation, and in terms of time as growth waves.

Fig. 5.6.14. Completed time-dependence of ring formation for the system shown in the insert. Gel precharged ($A_G = 1$) to $X = 75$; $X = N\Delta X$, here with $\Delta X = 1$, $L = 100$. Dimensionless variables. After Henisch and García-Ruiz (1986b).

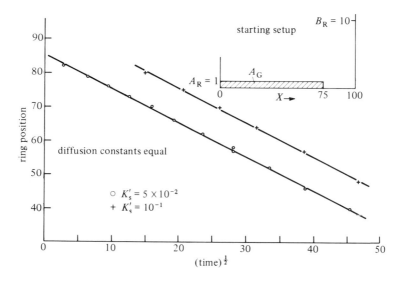

Fig. 5.6.15. Computer growth waves. Total mass at location $X = 25$, as a function of time (in computer units). System length: $L = 50$. $A_R = 100$, $A_G = 10$, growth factor $G_F = 5 \times 10^{-7}$, $R_{DC} = 10^{-4}$.

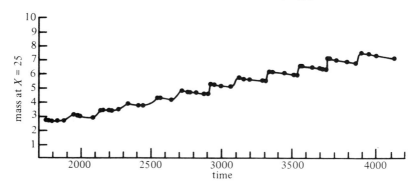

Cautions and opportunities

Though it is in the nature of computation to yield *numbers*, the significance of this work remains largely qualitative. This is because in order to make (even) the computer analysis tractable, important simplifying assumptions have to be made. One is total and instant ionization of the reacting species. Thus, the diffusing (A) and (B) species are seen as ions, forming neutral (AB) when they come together in a precipitate, *but not otherwise*. No provision is made for the diffusion of (AB) as such. Similarly, no provision at all has been made for the sometimes very sensitive effects which accidental trace impurities can have on the precipitation process; e.g. see Foster (1918, 1919), Dogadkin (1928), Gnanam *et al.* (1980), Ambrose *et al.* (1983b, c), and Krishnan *et al.* (1981). In this connection, two works with congenial culinary connections compete for attention, one on the effect of onion juice on Liesegang Rings (Stempell, 1929), the other on the effect of garlic (Siebert, 1930). The effect of dry Vermouth remains to be explored. It is also assumed by the computation that all local (A), for instance, is freely available, and that none is adsorbed on cell walls, or on the grains of previously formed precipitates as envisaged by Kanniah *et al.* (1981a, b). Accordingly, it was taken for granted that no additional rate processes are involved in solute capture and release. Though such processes have not been given great prominence in recent work, there was a time, e.g. see Bradford (1916, 1917, 1920a, b, 1921, 1922), when they were regarded as crucial. Indeed, Bradford ascribed the entire process of ring formation to the adsorption of reagents on precipitates. The fact that prominent rings can exist, consisting of only a few well-developed crystals (necessarily of small total surface area) makes this notion implausible as an operative mechanism, but it may still play a subsidiary role. Such a role would depend, of course, on the nature of the gel material. In connection with gelatin (not our favorite medium here), it has been the subject of intensive investigations; see, for instance, Van Hook (1941a, b), and Kruyt and Boelman (1932). Moreover, one cannot ignore Kirov's (1980) observation, according to which gel systems sometimes support the precipitation of more than one phase; different rate constants would then be expected.

Another possible complication was envisaged by Bechhold (1905), who pointed out that the solubility of a precipitate is not constant, but variable with composition of the surrounding liquid. He too maintained, as did Ostwald Wo. (1925) after him, that this could have important consequences for ring formation. Indeed, these relationships are in addition to those envisaged by Flicker and Ross (1974) and Ortoleva (1984) relating to the differential solubility of large and small grains, which has been advanced (see Section 5.4) as a potentially autonomous cause of periodic precipitation.

Lastly, the diffusion coefficients are taken as constants throughout, unmodified (even locally) by the formation of precipitates, no matter how dense and disruptive of the gel medium such precipitates might be. For computations, this is a matter of great expediency, but its validity should not be taken for granted. Indeed, the literature abounds with models that ascribe a primary function to the modification of diffusion constants by the precipitates themselves, e.g. see Fricke (1923), Traube and Takehara (1924), and Fischer and McLaughlin (1922). Packter (1956a, b) has also demonstrated that the structure of gelatin may be influenced by the reagents themselves and by substances otherwise present, which may have no role in the deposit formation as such. In particular, formaldehyde and detergents were found to have an influence on the formation of silver chromate rings, which was ascribed to this effect.

In view of all these uncertainties, it is not surprising that Van Hook (1938a, b), described the whole subject as 'highly controversial', even though its mysteries had by then been under investigation for almost half a century. He nevertheless provided a good deal of cogent evidence for the applicability of the supersaturation model to many systems, and for its essential correctness. In this respect he confirmed earlier work by Bolam (1928, 1930, 1933) on silver chromate and lead iodide. However, beyond applicability and 'essential correctness'', there are important details wich remain to be clarified. Their practical impact on computational analysis will be a matter of future research. The very necessity of supersaturation as a precipitation condition has been challenged (e.g. see Flicker and Ross, 1974) even in the context of systems to which it is widely believed to apply, and it is in any event clear that the counter-diffusion-supersaturation model cannot apply to such cases as, for instance, Liesegang Ring formation in melts or rhythmic crystallization from solution. McMasters and co-workers (1935) have interpreted such manifestations of ring growth in terms of thermal effects. They argued that local precipitation would be at least in part adiabatic, and should therefore lead to temperature gradients. The notion is that, as precipitation continues, the local temperature rises, 'until a condition is reached when there is a region immediately preceding the crystallization wave front that is no longer supercooled'. This region would correspond to the spacing of the rings; it must be 'traversed before crystallization begins anew' (Van Hook, 1938a, b); see also Nutting (1929). In contrast, the conditions envisaged above are always isothermal. Needless to say, they also make no allowance for gravity effects of the kind revealed by Kai *et al.* (1982), for the very best reason in the world: we do not know how. Yet another model, that involving 'competitive particle growth', based on the differential solubility of large and small precipitate grains, has already been mentioned in Section 5.4; see also Ortoleva, (1984).

It would no doubt be possible to devise algorithms which encompass all these complications, including those arising from thermal factors, but this has not yet been done. Meanwhile, even the small selection of 'experiments' described above is sufficient to show that a computer analysis can clarify and elucidate many of the complex phenomena that come under the heading of Liesegang Ring formation.

Availability of computer software

A programme-disk ('IN VITRO VERITAS') for IBM-compatible microcomputers is available from The Carnation Press, PO Box 101, State College, Pennsylvania, 16804, USA. It supports a great variety of hypothetical experiments on Liesegang Ring formation and growth waves, with parameters set by the user. The results may be graphically displayed in color or monochrome , or else printed out. Documentation included.

APPENDIX 1

Supplementary notes on materials grown in gels

Substance	Notes	Reference
Ammonium dihydrogen phosphate (ADP)	Good, transparent crystals up to $30 \times 6 \times 5$ mm.	Shanmugham *et al.* (1986)
	Crystal growth and doping.	Gits *et al.* (1985)
Ammonium tartrate	Dendrites, grown in silica gel. Correlations of growth features and growing conditions.	Saraf *et al.* (1986b)
Barium carbonate		Pietro *et al.* (1980)
Barium molybdate		Cho *et al.* (1977)
Cadmium tetraiodide	This greenish-black compound is unstable and ultimately decomposes into CdI_2.	Trebbe and Plewa (1982)
Calcium hydrogen phosphate (brushite)	Size up to 15 mm length; 'near-perfect quality'. A biologically important calcium phosphate.	Lefaucheux *et al.* (1979)
	See also	Legeros and Legeros (1972)
	and	Patel and Arora (1973)
Calcium oxalate	Growth in gel after ion exchange.	Cody *et al.* (1982)
Calcium sulfate (gypsum)	Needle-shaped crystals, about 1 mm diameter and several millimeters long. Comparison between silica gel and agar as growth media.	Van Rosmalen *et al.* (1976)
Cobalt oxinate	Liesegang Rings.	Kanniah *et al.* (1981b)
Copper	Perfect tetrahedra obtained, using a 1% hydroxylamine hydrochloride solution to reduce copper sulfate incorporated in the gel.	Holmes (1926)
	See also	Arora (1981)
Fluoroperovskites $(K, Rb, NH_4, Tl, Cs, Mn, Fe)$	Sizes up to ~0.7 mm.	Leckebusch (1974)
Gadolinium tartrate	Spherical crystals grown in silica gel; study of growth parameters, including pH and gel age.	Kotru *et al.* (1986)

Substance	Notes	Reference
Gold	Some of earliest records: reduction of gold salts, using oxalic acid, ammonium formate, ferrous sulfate, carbon monoxide, sulfur dioxide, hydrogen and ethylene.	Hatschek and Simon (1912a,b)
	Preparation also described: Liesegang rings produced, with a few larger crystals scattered between them.	Holmes (1926)
	See also	Kratochvil *et al.* (1968)
Iodine	Description of method.	Miller (1937a,b)
Lead	Dendrites. Gel contained lead acetate and metallic zinc.	Simon (1913)
	Experimental details.	Holmes (1926)
Lead carbonate	Ammonium carbonate and lead nitrate used as reagents. Whiskers.	Pillai and Ittyachen (1978)
Lead halide (PbCl$_2$, PbBr$_2$)	Needles of 15 mm length obtained in silica gels, shorter in gelatin and agar.	Hatschek (1911)
	See also	Abdulkhadar and Ittyachen (1982)
Lead hydrogen arsenate	Fine crystals, up to $26 \times 1 \times 4 \, mm^3$ and useful crystals up $20 \times 1.5 \times 8 \, mm^3$. See Fig. 2.1.1	Březina and Havránková (1980)
Lead hydrogen phosphate	Material interesting because it undergoes a ferroelectric–paraelectric phase transition. Description of method and etch pit studies.	Pandya *et al.* (1986)
	See also	Březina *et al.* (1975) Březina and Horváth (1980) Březina and Horváth (1981) Horváth (1983) Lefaucheux *et al.* (1986)
	Reaction of lead nitrate and phosphoric acid in a cross-linked polycrylanide gel.	Březina *et al.* (1976)
Lead nitrate-orthophosphate dihydrate	Good platelets of several mm size.	Březina and Havránková (1976)
Lead tartrate	Lead nitrate used as lead source; dendrite growth in silica gel. Sizes up to 2 cm.	Abdulkhadar and Ittyachen (1977)
Magnesium hydroxide	Liesegang Rings in agar.	Ambrose *et al.* (1982a, b, 1983a)
Manganese carbonate	Calcium and nickel ions incorporated as in up to $5 \times 10^{-2} \, M$ amounts. Habit modification due to doping noted. Manganese chloride and ammonium carbonate used as reagents.	Franke *et al.* (1979)

Substance	Notes	Reference
Mercuric chloride	$HgCl_2 \cdot 2H_2O$. Investigation of factors which control results. Crystals up to 2.5 cm in size.	Kurz (1966a)
Mercuric iodide	Mercuric chloride diffused into silica gel containing potassium iodide. Needle-shaped crystals, sometimes yellow at first, then red.	Holmes (1917)
	See also	Miller (1937a)
	More recent work.	Kurz (1966a)
Phosphates of nickel and cobalt	Crystals approaching 100 μm in diameter. Dissimilar results, despite the frequently isomorphous nature of nickel and cobalt salts.	Kurz (1966b)
Potassium dihydrogen phosphate (KDP)	Good crystals up to 1 cm in size. Good, transparent crystals up to $30 \times 6 \times 5 \ mm^3$	Březina and Havránková (1971) Shanmugham et al. (1986)
Potassium perchlorate	Studies of nucleation density. Crystals up to 5 mm or so in size. Aged gels supported less nucleation than new ones.	Patel and Rao (1977, 1978)
Selenium	Single Crystals reported.	Blank et al. (1968)
Silver periodates	Use of hybrid method Liesegang Ring pattern superimposed on macroscopic growth. Fine yellow crystals, up to 250 mm^3 in volume.	Arend and Huber (1972)
	See also	Arend and Perison (1971)
Silver sulfate	Very fast growth in water-glass containing 1.5 M sulfuric acid, with 1 M silver nitrate solution supernatant.	Holmes (1926)
Thallium halides (various)	Growth in gelatin.	Hausmann (1904)
Thallium iodide	Reaction of sodium iodide and thallium nitrate. β phase grows first, later develops into α phase, but without change of shape. Size up to 1.5 mm.	Shiojiri et al. (1978)
	See also	Levy and Mooser (1972)
Zeolites	Sizes up to 100 μm grown in acrylic acid/polymer. Aqueous slurries of sodium aluminate on one side and sodium metasilicate on the other served as reagents.	Ciric (1967)
	See also Fig. A.1.1 and	Charnell (1971) Joshi and Bosker (1979)
	Crystals in fiber form (21 mm), using thiethanolamine as stabilizing and complexing agent. Identification by infrared spectroscopy.	
Zeolites cyanoferrates (−hexa)	Structure determination Ion exchange behavior.	Gravereau et al. (1979) Garnier et al. (1982)

Fig. A.1.1. Zeolite crystals grown in silica gel. (*a*)
$Na_2O \cdot Al_2O_3 \cdot 2SiO_2 \cdot XH_2O$ (cubes) (*b*) $Na_2O \cdot Al_2O_3 \cdot 2.8SiO_2 \cdot XH_2O$
(octahedra) (Contributed by J. F. Charnell.)

(*a*)

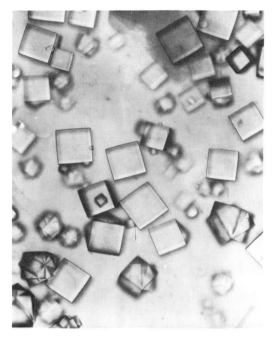

(*b*)

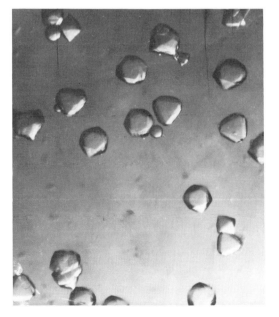

APPENDIX 2

Measurements on gel-grown crystals

No comprehensive survey is available, but some typical examples can be cited.

Substances	Measurements	References
Ammonium dihydrogen phosphate (ADP)	Electrical conductivity and micro-hardness.	Shanmugham *et al.* (1986)
Ammonium fluoroperovskite	Optical absorption spectrum.	Leckebusch (1974)
Ammonium tartrate	Study of etch patterns.	Saraf *et al.* (1986a)
Cadmium oxalate	Lattice constants.	Bridle and Lomer (1965)
	Surface features.	Cody *et al.* (1982)
Calcium tartrate	Electronic spin resonance.	Wakim *et al.* (1965)
Cesium–zinc (hexacyanoferrates)	Crystal structure determination	Gravereau and Garnier (1983) Gravereau *et al.* (1979)
Copper tartrate	Surface features.	Cody *et al.* (1982)
Gadolinium tartrate	SEM studies of morphology.	Kotru *et al.* (1986)
Lead	Lattice imperfections.	Bedarida (1964)
Lead hydrogen phosphate	Morphology and imperfections.	Březina and Havránková (1980)
Lead iodide	Raman spectra and photo-decomposition.	Khilji *et al.* (1982a, b)
	Crystal studies; polytypism.	Chand and Triguanyai (1976)
	Band and defect level structure.	Henisch and Srinivasagopalam (1966) Dugan and Henisch (1967a) Dugan and Henisch (1967b) Dugan and Henisch (1968)
	Crystal structure; polytypism.	Mitchell (1959) Hanoka *et al.* (1967) Pinsker *et al.* (1943)
Manganese carbonate	Electron microscope studies of crystal habit.	Franke *et al.* (1979)

Substances	Measurements	References
Potassium dihydrogen phosphate (KDP):		
pure	Electrical conductivity parallel to c-axis and perpendicular to it.	Shanmugham *et al.* (1982)
	Microhardness.	Shanmugham *et al.* (1986)
	Defect structure.	Gnanam *et al.* (1981)
oxalate doped	Electrical conductivity.	Shanmugham *et al.* (1985)
Silver iodide	Optical properties.	Suri and Henisch (1971)
	Ionic conductivity.	Cochrane and Fletcher (1971)
	Optical and electrical properties.	Govindacharyulu and Bose (1977)
	Drift mobility.	
	Photoconductivity.	Bose and Govindacharyulu (1976)
	DC polarization.	Mazumdar *et al.* (1982)
	Proof of crystallographic polarity.	Bhalla *et al.* (1971)
	Ionic DC conductivity.	Gon (1986)
Tetramethylammonium tetrachlorozincate	Synchotron x-ray topography	Ribet *et al.* (1986)
Tin iodides	Electrical conductivity; magnetic susceptibility	Desai and Rai (1981, 1984)

REFERENCES

The numbers following each item refer to sections in this book

Abdulkhadar, M. and Ittyachen, M. A. (1977) *J. Cryst. Growth* **39**, 365   3.6, 4.1, A.1

Abdulkhadar, M. and Ittyachen, M. A. (1980) *Proc. Ind. Acad. Sci.* **89**, 69   4.1

Abdulkhadar, M. and Ittyachen, M. A. (1982) *Cryst. Res. and Technol.* **17**, 33   3.6, A.1

Adair, G. S. (1920) *Biochem. J.* **14**, 762   2.1, 5.4

Adair, J. (1986) Personal communication   5.3

Alexander, A. E. and Johnson, P. (1949) *Colloid Science*, Clarendon Press, Oxford   2.1

Alexander, G. B. (1953) *J. Amer. Chem. Soc.* **75**, 5655   2.2

Alexander, G. B. (1954) *J. Amer. Chem. Soc.* **76**, 2094   2.2

Alexander, G. B. (1967) *Silica and Me*, Anchor Books Doubleday, Garden City, New Jersey   2.2

Ambrose, S., Gnanam, F. D. and Ramasamy, P. (1983a) *Proc. Indian Acad. Sci. (Chem. Soc.)* **92**, 239   A.1

Ambrose, S., Gnanam, F. D. and Ramasamy, P. (1983b) *Cryst. Res. and Technol.* **18**, 1225   5.6

Ambrose, S., Gnanam, F. D. and Ramasamy, P. (1983c) *Cryst. Res. and Technol.* **18**, 1231   5.6

Ambrose, S., Gnanam, F. D. and Ramasamy, P. (1984) *J. Colloid and Interface Sci.* **97**, 296   5.3

Ambrose, S., Kanniah, N., Gnanam, F. D. and Ramasamy, P. (1982a) *Cryst. Res. and Technol.* **17**, 299   A.1

Ambrose, S., Kanniah, N., Gnanam, F. D. and Ramasamy, P. (1982b) *Cryst. Res. and Technol.* **17**, 609   A.1

Amorós, J. L. and García-Ruiz, J. M. (1982) *Investigación y Ciencia* 6W:71 (August)   3.6

Amsler J. and Scherer P. (1941) Helv. Phys. Acta, **14**, 318   4.1

Amsler, J. and Scherer P. (1942) Helv. Phys. Acta, **15**, 699   4.1

Anderson, J. S. (1914) *Z. Phys. Chem.* **20**, 191   2.2

Aragón, P., Santos, M. J., Maleiraa, A., Gallego-Andrew, R. and García-Ruiz, J. M. (1984) *An. Quim.* **80**, 134   3.6

Arend, H. (1986) Personal Communication   1.3

Arend, H. and Huber, W. (1972) *J. Cryst. Growth* **12**, 179   3.5, A.1

Arend, H. and Perison, J. (1971) *Mat. Res. Bull.* **6**, 1205   3.5, A.1

Armington, A. F., Dipietro, M. A. and O'Connor, J. J. (1967a) *Air Force Cambridge Research Laboratories (Ref.* 67-0445). *Physical Science Research Paper No.* 334 *(July)*   1.3, 1.4, 3.1, 4.5

Armington, A. F., Dipietro, M. A. and O'Connor, J. J. (1967b) *Air Force Cambridge Research Laboratories (Ref.* 67-0304). *Physical Sciences Research Paper No.* 325 *(May)*   1.3, 1.4, 3.1

Armington, A. F. and O'Connor, J. J. (1967) *Mat. Res. Bull.* **2**, 907   1.3

Armington, A. F. and O'Connor, J. J. (1968a) *Proc. Int. Conf. Crystal Growth*, Birmingham University, UK, July 15-19, 1968; *J. Cryst. Growth* **3-4**, 367   1.3

Armington, A. F. and O'Connor, J. J. (1968b) *Mat. Res. Bull.* **3**, 923   1.3

Arora, S. K. (1981) *Proc. Cryst. Growth and Charact.* **4**, 345   1.3, 4.1, A.1

Arredondo Reyna, L. M. L. (1980) *Determinación de Coefficientes de Diffusión en Geles*, Chemistry Thesis. Universidad Nacional Anbinoma de México   2.3

Aschoff, L. (1939) *Kolloid Z.* **89**, 107   1.1

Audsley, A. and Aveston, J. (1962) *J. Amer. Chem. Soc.* **84**, 2320   2.2

Bajpai, A. C., Calus, I. M. and Fairley, J. A. (1977) *Numerical Methods for Engineers and Scientists* John Wiley and Sons, London   2.3

Banbury, P. C., Gebbie, H. A. and Hogarth, C. A. (1951) *Semiconducting Materials* (Editor, H. H. Henisch) Butterworths, London   3.6

Barber, P. G. (1986) Personal communication   2.1

Barber, P. G. and Simpson, N. R. (1985) *J. Cryst. Growth* **73**, 400   2.1

Bárta, C. and Žemlička, J. (1967) Personal communication (Institute of Solid State Physics, Prague)   3.5

Bartell, F. E. (1936) *Laboratory Manual of Colloid and Surface Chemistry*, Edwards Brothers, Ann Arbor, Michigan   2.1

Bechhold, H. (1905) *Z. phys. Chem.* **52**, 185   5.6

Bechhold, H. and Zeigler, K. (1906) *Ann. Phys.* **20**, 900   2.1

Beck, C. W. and Bender, M. J. (1969) *J. Urology* **101**, 208   3.5

Becker, R. (1938) *Ann. Phys.* **32**, 128   4.1

Becker, R. (1940) *Proc. Phys. Soc.* **52**, 70   4.1

Bedarida, F. (1964) *Periodico di Mineralogia* **33**, 1   A.2

Beekley, J. S. and Taylor, H. S. (1925) *J. Phys. Chem.* **29**, 942   5.4

Bchm, H. (1984) Unpublished. personal communication   5.3

Berg, W. F. (1938) *Proc. Roy. Soc. A.* **164**, 79   3.1, 3.4

Bernard, Y., Lefaucheux, F., Gits, S., Robert, M. C. and Picard, M. C. (1982) *C.R. Acad. Sci.* (Paris) **295**, 1065   3.1

Bernard, Y., Robert, M. C. and Lefaucheux, F. (1985) *C.R. Acad. Sci.* (Paris) **301**, 1105   3.1

Bhalla, A. S., Suri, S. K. and White, E. W. (1971) *J. Appl. Phys.* **42**, 1835   1.3

Blank, Z. and Brenner, W. (1971) *J. Cryst. Growth* **11**, 258   1.3, 3.6

Blank, Z., Brenner, W. and Okamoto, Y. (1968) *Mat. Res. Bull.* **3**, 555   3.6

Blank, Z., Speyer, D. M., Brenner, W. and Okamoto, Y. (1967) *Nature* **216**, 1103   1.3, A.1

Blitz, M. (1927) *Z. phys. Chem.* **126**, 356   2.2

Böhm, J. (1927) *Kolloid Z.* **42**, 276   2.2

Bolam, T. R. (1928) *Trans. Faraday Soc.* **24**, 463   4.3, 5.6

Bolam, T. R. (1930) *Trans. Faraday Soc.* **26**, 133   4.3, 5.6

Bolam, T. R. (1933) *Trans. Faraday Soc.* **29**, 864   4.3, 5.6

Bolam, T. R. and Mackenzie, M. R. (1926) *Trans Faraday Soc.* **22**, 151 and 162   4.3

Bose, D. N. and Govindacharyulu, P. A. (1976) *Proc. 13th Int. Conf. on Phys. of Semiconductors, Rome*, North Holland Publishing Co. Amsterdam, p. 1180   A.2

Bradford, S. C. (1916) *Biochem. J.* **10**, 169   3.2, 5.4, 5.6

Bradford, S. C. (1917) *Biochem. J.* **11**, 157   3.2, 5.6

Bradford, S. C. (1920a) *Biochem. J.* **14**, 29   5.6

Bradford, S. C. (1920b) *Biochem. J.* **14**, 474   5.6

Bradford, S. C. (1921) *Science* **54**, 463   5.6

Bradford, S. C. (1922) *Kolloid Z.* **30**, 364   5.6

Bradford, S. C. (1926) *Colloid Chemistry* (Editor, J. Alexander), Chemical Catalog Co., New York   1.1, 5.4

Brenner, W., Blank, Z. and Okamoto, Y. (1966) *Nature* **212**, 392   1.3, 3.6

Březina, B. and Havránková, M. (1971) *Mat. Res. Bull.* **6**, 537   A.1

184   *References*

Březina, B. and Havránková, M. (1976) *J. Cryst. Growth* **34**, 248   2.1, A.1

Březina, B. and Havránková, M. (1980) *Kristall und Technik.* **15**, 1447   2.1, A.1, A.2

Březina, B., Havránková, M. and Dušek, K. (1976) *J. Cryst. Growth* **34**, 248   3.1, A.1

Březina, B. and Horváth, J. (1981) *J. Cryst. Growth* **52**, 858   2.1, 3.2, A.1

Březina, B., Smutný, F. and Fousek, J. (1975) *Czech. J. Phys.* **25**, 1411   A.1

Bridle, C. and Lomer, T. R. (1965) *Acta Cryst.* **19**, 483   A.2

Brummer, E. (1904) *Z. phys. Chem.* **47**, 56   3.1

Buchholcz, J. (1941) *Kolloid Z.* **96**, 72   5.2

Buckley, H. A. (1952) *Crystal Growth*, John Wiley and Sons, New York   4.1

Bulger, G. (1969) Personal communication (The Pennsylvania State University)   3.3

Burton, E. F. and Bell, X. (1921) *J. Phys. Chem.* **25**, 526   5.2

Burton, W. K., Cabrera, N. and Frank, F. C. (1951) *Phil. Trans. Roy. Soc.* **243**, 299   3.6, 4.1

Caslavska, V. and Gron, P. (1984) *Caries Research* **18**, 354   2.1

Chand, M. and Triguanyai, G. C. (1976) *J. Cryst. Growth* **35**, 307   A.1

Charnell, J. F. (1971) *J. Cryst. Growth* **8**, 291   A.1

Chatterji, A. C. and Rastogi, M. C. (1951) *J. Ind. Chem. Soc.* **28**, 128   1.3, 5.4

Cho, S. A., Gomez, J. A., Camisotti, R. and Ohep, J. C. (1977) *J. Mat. Sci.* **12**, 816   A.1

Christomanos, A. (1950) *Nature* **N:4189**, 238   1.3

Ciric, J. (1967) *Science* **155**, 689   A.1

Coatman, R. D., Thomas, N. L. and Double, D. D. (1980) *J. Mat. Sci.* **15**, 217   3.6

Cochrane, G. (1967) *Brit. J. Appl. Phys.* **18**, 687   1.3

Cochrane, G. and Fletcher, N. H. (1971) *J. Phys. Chem. Solids* **32**, 2557   A.2

Cody, A. M., Horner, H. T. and Cody, R. D. (1982) in *Scanning Electron Microscopy*, SEM Inc., AMF O'Hare, Chicago, Illinois   2.1, A.1, A.2

Copisarow, M. (1931) *Kolloid Z.* **54**, 257   5.2

Cornu, A. (1909) *Kolloid Z.* **4**, 5   1.1

Crank, J. (1956) *Mathematics of Diffusion*, Oxford University Press, Oxford   3.1

Dancy, E. A. (1969) Westinghouse Research Laboratories, Personal Communication   1.3

Daus, W. and Tower, O. F. (1929) *J. Phys. Chem.* **33**, 605   5.2

Davies, E. (1923) *J. Amer. Chem. Soc.* **45**, 2261   1.2

Davies, E. C. H. (1917) *J. Amer. Chem. Soc.* **39**, 1312   5.2

Dee, G. T. (1986) *Phys. Rev. Lett.* **57**, 275   5.6

DeHaas, Y. M. (1963) *Nature* **200**, 876   1.3

Deiss, E. (1939) *Kolloid Z.* **89**, 146   1.1

Dekeyser, W. L. and Degueldre, L. (1950) *Bull. Soc. Chim. Belg.* **49**, 40   3.5

Dennis, J. (1967) *Crystal Growth in Gels.* Thesis. The Pennsylvania State University   3.1, 3.2, 3.3, 4.5

Dennis, J. and Henisch, H. K. (1967) *J. Eletrochem. Soc.* **114**, 263   1.3, 3.3, 4.4

Dennis, J., Henisch, H. K. and Cherin, P. (1965) *J. Electrochem. Soc.* **112**, 1240   1.3

Deryagin, B., Friedlyand, R. F. and Krylova, V. (1948) *Doklady Akad. Nauk. SSSR* **61**, 653   2.2

Desai, C. C. and Rai, J. L. (1981) *J. Cryst. Growth* **53**, 432   1.3

Desai, C. C. and Rai, J. L. (1984) *Surface Technol.* **22**, 189   1.3

Dhanasekaran, R. and Ramasay, P. (1981a) *Cryst. Res. and Technol.* **16**, 635   3.1

Dhanasekaran, R. and Ramasay, P. (1981b) *Cryst. Res. and Technol.* **16**, 863   3.1

Dhanasekaran, R. and Ramasay, P. (1982) *J. Phys. D: Appl. Phys.* **15**, 1047   3.1

Dhar, N. R. and Chatterji, A. C. (1922) *Kolloid Z.* **31**, 15   4.1

Dhar, N. R. and Chatterji, A. C. (1924) *J. Phys. Chem.* **28**, 41   5.2

Dhar, N. R. and Chatterji, A. C. (1925a) *Kolloid Z.* **37**, 2   1.3, 5.2, 5.4

Dhar, N. R. and Chatterji, A. C. (1925b) *Kolloid Z.* **37**, 89   4.3, 5.2

Dhar, N. R. and Chatterji, A. C. (1934) *J. Phys. Chem.* **28**, 41   5.2

Dogadkin, B. (1928) *Kolloid Z.* **40**, 33   5.6

Dreaper, G. (1913) *J. Soc. Chem. Ind.* **32**, 678   1.2

Dugan, A. E. (1967) The Pennsylvania State University Personal communication 1.3

Dugan, A. E. and Henisch, H. K. (1967a) *J. Phys. Chem. Solids* **28**, 1885 A.2

Dugan, A. E. and Henisch, H. K. (1967b) *J. Phys. Chem. Solids* **28**, 971 A.2

Dugan, A. E. and Henisch, H. K. (1968) *Phys. Rev.* **171**, 1047 A.2

Dundon, M. L. and Mack, E. (1923) *J. Amer. Chem. Soc.* **45**, 2479 4.1

Dunning, W. J. (1955) Chemistry of the Solid State (Editor W. E. Garner), Butterworths, London 4.1

Dunning, W. J. (1973) *Particles Growth in Suspensions* (Editor, A. Smith), Academic Press, New York 5.4

Egli, P. H. and Johnson, L. R. (1963) *The Art of Crystal Growing* (Editor, J. J. Gilman), John Wiley & Sons, New York 3.1

Eitel, W. (1954) *The Physical Chemistry of Silicates*, The University of Chicago, Chicago 1.1, 2.1, 2.2

Endres, H. A. (1926) *Colloid Chemistry* (Editor, J. Alexander), Chemical Catalog Co., New York 1.1

Evans, G. J. (1984) *Mat. Lett.* **2**, 420 1.3

Everett, D. H. (1973) *Colloid Sciences*, Vol. 1, Chemical Society, London 5.4

Faust, J. W. (1968) The Pennsylvania State University Personal Communicaion 3.1, 4.5

Feinn, D., Ortoleva, P., Scalf, W., Schmidt, S. and Wolff, M. (1978) *J. Chem. Phys.* **67**, 27 5.4

Fife, P. C. (1976) *J. Chem. Phys.* **64**, 554 5.4

Fischer, M. H. and McLaughlin, G. D. (1922) *Kolloid Z.* **30**, 13 5.6

Fisher, L. W. (1928) *Amer. J. Sci.* **15**, 39 1.3

Fisher, L. W. and Simons, F. L. (1926a) *Amer. Mineralogist* **11**, 124 1.1

Fisher, L. W. and Simons, F. L. (1926b) *Amer. Mineralogist* **11**, 200 1.3, 3.5

Flicker, M. and Ross, J. (1974) *J. Chem. Phys.* **60**, 3458 3.2, 5.4, 5.6

Foster, A. W. (1918) *Trans. Roy. Soc. Can.* **12**, 55 4.3, 5.6

Foster, A. W. (1919) *J. Phys. Chem.* **23**, 645 5.6

Foster, J. (1919) *J. Phys. Chem.* **23**, 145 5.4

Frank, F. C. (1949) *J. Faraday Soc.* **5**, 48 4.1

Frank, F. C. (1950) *Proc. Roy. Soc.* **201A**, 586 3.1

Frank, F. C. (1951a) *Phil. Mag.* **42**, 1014 4.1, 4.3

Frank, F. C. (1951b) *Acta Cryst.* **4**, 497 4.3

Franke, W., Ittyachen, M. A. and Pillai, K. M. (1979) *Promāna, India* **13**, 293 A.1, A.2

Franz, V. (1910) *Arch. f. Vergl. Optothalmogie* **1**, 283 5.2

Fricke, R. (1923) *Z. phys. Chem.* **107**, 41 5.6

Fricke, R. and Suwelack, O. (1926) *Z. phys. Chem.* **124**, 359 2.1, 5.3

Füchtbauer, C. (1904) *Z. phys. Chem.* **48**, 566 5.2

García-Ruiz, J. M. (1982) *Crystal Growth in Semiconducting Environments* (Editors, R. Rodriguez and I. Sunagawa), Estudios Geol. **38**, 209 3.6

García-Ruiz, J. M. (1986) *J. Cryst. Growth* **75**, 441 3.6

García-Ruiz, J. M. and Miguez, F. (1982) *Estudios Geol.* **38**, 3 5.5

García-Ruiz, J. M., Santos, A and Alfaro, E. J. (1987). Personal communication 5.6

Garnier, E., Gravereau, P. and Hardy, A. (1982) *Acta Cryst.* **B38**, 410 A.1

Gebhardt, W. (1912) *Verk. d.d. Zool. Ges.* **22**, 175 1.1, 5.2

George, M. T. (1986) Personal communication 3.1

George, M. T. and Vaidyan, V. K. (1981a) *J. Cryst. Growth* **53**, 300 1.3, 3.1

George, M. T. and Vaidyan, V. K. (1981b) *J. Appl. Cryst.* **14**, 345 1.3, 3.1

George, M. T. and Vaidyan, V. K. (1982a) *Cryst. Res. and Technol.* **17**, 313 1.3, 3.1

George, M. T. and Vaidyan, V. K. (1982b) *J. Appl. Electrochem.* **12**, 359 1.3, 3.1

Gerrard, J. E., Hoch, M. and Meeks, F. R. (1962) *Acta Metallurgica* **10**, 751 3.2, 5.2, 5.4, 5.5

Ghosh, J. (1930) *J. Ind. Chem. Soc.* **7**, 509 4.3

Gits, S., Robert, M. C. and Lefaucheux, F. (1985) *J. Cryst. Growth* **71**, 203 A.1

Glansdorff, P. and Prigogine, I. (1971) *Thermodynamics of Structure, Stability and Fluctuations*, Interscience, New York   5.4

Glocker, D. A. and Soest, I. F. (1969) *J. Chem. Phys.* **51**, 3143   1.3

Gnanam, F. D., Krishnan, S., Ramasamy, P. and Laddha, G. S. (1980) *J. Colloid and Interface Sci.* **73**, 193   3.5, 5.2, 5.6

Gnanam, F. D., Ramasamy, P. and Somasundram, M. (1981) *Ind. J. of Pure and Appl. Phys.* **19**, 1105   A.1

Gon, H. B. (1985) *Nat. Acad. Sci. (Ind.) Lett.* **8**, 51   1.3

Gon, H. B. (1986) *Phys. Stat. Sol.* **(a)94**, K61   A.2

Gore, V. (1936a) *Kolloid Z.* **76**, 193   3.2

Gore, V. (1936b) *Kolloid Z.* **76**, 330   3.2

Gore, V. (1938a) *Kolloid Z.* **82**, 79   3.2

Gore, V. (1938b) *Kolloid Z.* **82**, 203   3.2

Govindacharyulu, P. A. and Bose, D. N. (1977) *J. Appl. Phys.* **48**, 1381   A.2

Graham, T. (1862) *Liebig's Ann.* **121**, 5   2.1

Gravereau, P. and Garnier, E. (1983) *Revue de Chimie Minérale* **20**, 68   A.2

Gravereau, P., Garnier, E. and Hardy, A. (1979) *Acta Cryst.* **B35**, 2843   A.1, A.2

Greenberg, S. A. and Sinclair, D. (1955) *J. Phys. Chem.* **59**, 435   2.2

Gruzensky, P. M. (1967) *J. Phys. Chem. Solids* **Suppl. No. 1**, 365   3.5

Hahn, F. V. von (1925) *Kolloid Z.* **37**, 300   1.1

Halberstadt, E. S. (1967a) University of Reading, UK, Personal communication   1.3

Halberstadt, E. S. (1967b) *Nature* **216**, 574   1.3

Halberstadt, E. S. and Henisch, H. K. (1968) *J. Cryst. Growth*, **3/4**, 363   4.2, 4.5

Halberstadt, E. S., Henisch, H. K., Nickl, J. and White, E. W. (1969) *J. Colloid and Interface Sci.* **24**, 461   3.5, 4.4, 4.5

Hanoka, J. (1967) *Polytypism in PbI$_2$ and its Interpretation According to Epitaxial Theory*. Thesis. The Pennsylvania State University   4.3

Hanoka, J. (1968) Personal Communication   3.1, 3.4

Hanoka, J. I. (1969) *J. Appl. Phys.* **40**, 2694   3.4

Hanoka, J. I., Vedam, K. and Henisch, H. K. (1967) *Crystal Growth* (Editor, S. Peiser), Pergamon Press, New York; *J. Phys. Chem. Solids (Supplement)*, 369   A.2

Happel, P., Liesegang, R. E. and Mastbaum, O. (1929a) *Kolloid Z.* **48**, 80   1.3

Happel, P., Liesegang, R. E. and Mastbaum, O. (1929b) *Kolloid Z.* **48**, 252   1.3

Hatschek, E. (1911) *Kolloid Z.* **8**, 13   1.2, 3.5, A.1

Hatschek, E. (1914) *Kolloid Z.* **14**, 115   4.3, 5.2

Hatschek, E. (1919) *Brit. Assoc. Reports* **1919**, 23   4.3, 5.2

Hatschek, E. (1920) *Biochem J.* **14**, 418   5.2

Hatschek, E. (1921) *Proc. Roy. Soc.* **A99**, 496   4.5, 5.2

Hatschek, E. (1925) *Kolloid Z.* **37**, 297   4.5, 5.2

Hatschek, E. (1929) *Kolloid Z.* **49**, 244   2.2   ·

Hatschek, E. and Simon, A. L. (1912a) *J. Soc. Chem. Ind.* **31**, 439   A.1

Hatschek, E. and Simon, A. L. (1912b) *Kolloid Z.* **10**, 265   A.1

Hauser, E. A. (1955) *Silica Science*, Van Nostrand, Princeton, New Jersey   2.1, 2.2

Hausmann, J. (1904) *Z. anorg. Chem.* **40**, 110   A.1

Hedges, E. S. (1931) *Colloids*, Edward Arnold Co, London   1.1, 2.1

Hedges, E. S. (1932) *Liesegang Rings*, Chapman, London   1.2, 2.1, 5.2

Hedges, E. S. and Henley, R. V. (1928) *J. Chem. Soc.* **129**, 2714   2.1

Henisch, H. K. (1986) *J. Cryst. Growth* **76**, 279   5.5, 5.6

Henisch, H. K., Dennis, J. and Hanoka, J. I. (1965) *J. Phys. Chem. Solids* **26**, 493   1.3, 4.2, 4.3, 4.5

Henisch, H. K. and García-Ruiz, J. M. (1986a) *J. Cryst. Growth* **75**, 195   5.5, 5.6

Henisch, H. K. and García-Ruiz, J. M. (1986b) *J. Cryst. Growth* **75**, 203   5.5, 5.6

Henisch, H. K., Hanoka, J. I. and Dennis, J. (1965) *J. Electrochem. Soc.* **112**, 627   3.1, 4.2, 4.5

Henisch, H. K. and Srinivasagopalam, C. (1966) *Solid State Comm.* **4**, 415   A.2

Hermans, J. J. (1947) *J. Colloid. Sci.* **2**, 387   5.6

Higuchi, H. and Matuura, R. (1962) *Mem. Fac. Sci. Kyushu Univ. Ser.* **C5**, 33   5.3

Hirsch-Ayalon, P. (1957) *J. Polymer Sci.* **23**, 697   5.3

Holmes, H. N. (1917) *J. Phys. Chem.* **21**, 709   1.1, A.1

Holmes, H. N. (1918) *J. Amer. Chem. Soc.* **40**, 187   5.2, 5.3

Holmes, H. N. (1926) *Colloid Chemistry* (Editor, J. Alexander), Chemical Catalog Co, New York   1.2, A.1

Holtzapfel, L. (1942) *Kolloid Z.* **100**, 386   2.2

Horváth, J. (1983) *J. Appl. Phys.* **54**, 6749   A.1

Huber, W. (1959) *Z. angew. Math. Mech.* **19**, 1   3.1

Hurd, C. B. and Letteron, H. A. (1932) *J. Phys. Chem.* **36**, 604   2.2

Hurd, C. B., Raymond, C. L. and Miller, P. S. (1934) *J. Phys. Chem.* **38**, 663   2.2

Hurd, C. B. and Thompson, L. W. (1941) *J. Phys. Chem.* **45**, 1263   2.2

Huttel, K. P. (1984) *Fotogeschichte 4*, No. 12, 2   5.1

Ikornikova, N. Yu. and Butuzov, V. P. (1956) *Doklady Akad. Nauk, SSSR* **111**, 105   3.5

Iler, R. K. (1955) *The Colloid Chemistry of Silica and Silicates*, Cornell University Press, Ithaca, New York   2.2

Ives, M. B. and Plewes, J. (1965) *J. Chem. Phys.* **42**, 293   4.4

Jablczynski, K. (1923) *Bull. Soc. Chim.* (4) **33**, 1592   5.3

Jagodzinski, H. (1963) *Crystallography and Crystal Perfection* (Editor, G. N. Ramachandran), Academic Press, New York   3.1

Janek, A. (1923) *Kolloid Z.* **32**, 252   1.1, 5.2

Johnson, P. and Metcalf, C. J. (1963) *J. Photogr. Sci.* **11**, 214   2.1

Jones, W. J. (1913) *Z. phys. Chem.* **82**, 448   4.1

Jones, W. J. and Partington, J. R. (1915a) *J. Chem. Soc.* **103**, 1019   4.1

Jones, W. J. and Partington, J. R. (1915b) *Phil. Mag.* **29**, 35   4.1

Joshi, M. S. and Antony, A. V. (1980) *Bull. Mat. Sci.* **2**, 31   1.3

Joshi, M. S. and Boskar, B. T. (1979) *J. Cryst. Growth* **47**, 654   A.1

Kahlweit, M. (1960) *Z. phys. Chem.* **25**, 1   5.6

Kahlweit, M. (1962) *Z. phys. Chem.* **32**, 1   5.6

Kahlweit, M. (1975) *Adv. in Colloid and Interface Sci.* **5**, 1   5.6

Kai, S., Müller, S. C. and Ross, J. (1982) *J. Chem. Phys.* **76**, 1392   2.3, 5.3, 5.6

Kai, S. Müller, S. C. and Ross, J. (1983) *J. Chem. Phys.* **87**, 806   5.2, 5.6

Kanniah, N. (1983) *Revert and Direct Liesegang Phenomenon.* Thesis. Crystal Growth Center, Anna University, Madras, India   5.6

Kanniah, N., Gnanam, F. D., Ramasamy, P. and Laddha, G. S. (1981a) *J. Colloid and Interface Sci.* **80**, 369   5.4, 5.6

Kanniah, N., Gnanam, F. D. and Ramasamy, P. (1981b) *J. Colloid and Interface Sci.* **80**, 377   A.1

Kanniah, N., Gnanam, F. D. and Ramasamy, P. (1983) *J. Colloid and Interface Sci.* **94**, 412   5.3

Kanniah, N., Gnanam, F. D. and Ramasamy P. (1984) *Proc. Ind. Acad. Sci.* **93**, 801   5.3

Karrer, E. (1922) *J. Amer. Chem. Soc.* **44**, 951   5.2

Kasatkin, A. P. (1966) *Sov. Phys.* **11**, 295   4.5

Kaspar, J. (1959) *Growth of Crystals*, Consultants Bureau, New York   3.5

Keller, J. B. and Rubinow, S. I. (1981) *J. Chem. Phys.* **74**, 5000   5.6

Khaimov-Mal'kov, V.Ia. (1958) *Sov. Phys.* **3**, 487   3.2, 3.4

Khilji, M. Y., Sherman, W. F. and Wilkinson, G. R. (1982) *J. Raman Spectr.* **13**, 127   A.2

Kirov, G. K. (1968) *Kristall und Technik.* **3**, 573   3.1

Kirov, G. K. (1969) *Compt. Rend. Acad. Bulg. Sci.* **22**, 915   5.5

Kirov, G. K. (1972) *J. Cryst. Growth* **15**, 102   3.1, 3.5, 5.5, 5.6

Kirov, G. K. (1977) *Compt. Rend. Acad. Bulg. Sci.* **30**, 559   5.5
Kirov, G. K. (1978) *Compt. Rend. Acad. Bulg. Sci.* **31**, 449   2.3
Kirov, G. K. (1980) *Compt. Rend. Acad. Bulg. Sci.* **33**, 1659   5.5, 5.6
Kisch, B. (1929a) *Kolloid Z.* **49**, 154   1.3
Kisch, B. (1929b) *Kolloid Z.* **49**, 156   1.3
Kisch. B. (1933) *Kolloid Z.* **65**, 316   5.2
Kitano, Y. (1962) *Bull. Chem. Soc. Jap* **35**, 1980   3.5
Knapp, L. F. (1922) *Trans. Faraday Soc.* **17**, 457   4.1
Knöll, H. (1938a) *Kolloid Z.* **82**, 76   1.1, 5.2
Knöll, H. (1938b) *Kolloid Z.* **85**, 290   1.1
Knöll, H. (1939) *Kolloid Z.* **89**, 135   5.2
Koenig, A. E. (1920) *J. Phys. Chem.* **24**, 466   5.2
Köhler, F. (1916) *Kolloid Z.* **19**, 65   5.2
Köhn, M. and Mainzhausen, L. (1937) *Kolloid Z.* **79**, 316   4.5, 5.2
Koide, H. and Nakamura, T. (1943) *Proc. Imp. Acad. Jap.* **19**, 202   1.1
Köppen, R. (1936) *Z. angew Chem.* **228**, 169   4.1
Köppen, R. (1938) *Kolloid Z.* **89**, 219   2.1
Kotru, P. N., Gupta, N. K. and Raina, K. K. (1986) *Cryst. Res. and Tech.* **21**, 15   A.1, A.2
Kratochvil, P., Sprusil, B. and Heyrovsky, M. (1968) *J. Cryst. Growth* **3/4**, 360   4.2, A.1
Krejčí, L. and Ott, E. (1931) *J. Phys. Chem.* **35**, 2061   2.2
Krishnan, S., Gnanam, F. D., Ramasamy, P. and Laddha, G. S. (1981) *Cryst. Res. and Technol.* **16**, 1103   5.6
Krishnan, S., Gnanam, F. D., Ramasamy, P. and Laddha, G. S. (1982) *Cryst. Res. and Tech.* **17**, 307
Krusch, C. (1907) *Z. prakt. Geol.* **15**, 129   1.1
Krusch, C. (1910) *Z. prakt. Geol.* **18**, 165   1.1
Kruyt, H. R. (1952) *Colloid Science*, Elsevier, Amsterdam   5.4
Kruyt, H. R. and Boelman, X. (1932) *Kolloid Beihefte* **35**, 183   5.6
Kumagawa, M., Sunami, H., Terasaki, J. and Nishizawa, T. (1968) *Jap. J. Appl. Phys.* **7**, 1332   4.5
Kurbatov, V. (1931) *Z. Krist.* **77**, 164   5.5
Kurihara, H., Higuchi, H., Hirikawa, T. and Matuura R. (1962) *Bull. Chem. Soc. Jap.* **35**, 1740   2.1
Kurz, P. (1965) *Science and Technol. March issue*, 81   1.2
Kurz, P. (1966a) *Ohio J. Sci.* **66**, 284   A.1
Kurz, P. (1966b) *Ohio J. Sci.* **66**, 349   A.1
Küster, E. (1913) *Kolloid Z.* **13**, 192   5.2
Kuzmenko, S. M. (1928) *Ukraine Chem. J.* **3**, 231   5.2
Lakhani, M. P. and Mathur, R. N. (1934) *Kolloid Z.* **67**, 59   5.3
Leckebusch, R. (1974) *J. Cryst. Growth* **23**, 74   A.1, A.2
Lee, R. E. J. and Meeks, F. R. (1971) *J. Colloid and Interface Sci.* **35**, 584   3.2, 5.3, 5.4
Lefaucheaux, F., Robert, M. C. and Arend, H. (1979) *J. Cryst. Growth* **47**, 313   1.3, A.1
Lefaucheux, F., Robert, M. C., Bernard, Y. and Gits, S. (1984a) Personal communication   3.1, 3.4
Lefaucheux, F., Robert, M. C., Bernard, Y. and Gits, S. (1984b) *Cryst. Growth and Technol.* **19**, 1541   3.1, 3.4
Lefaucheux, F., Robert, M. C., Gits, S., Bernard, Y. and Gauthier-Manuel, B. (1986) *Rev. Int. Hautes Temp. Réfract.* **23**, 57   2.2
Lefaucheux, F., Robert, M. C. and Manghi, E. (1982) *J. Cryst. Growth* **56**, 141   3.1
Legeros, R. Z. and Legeros, J. P. (1972) *J. Cryst. Growth* **13/14**, 476   A.1
Lendvay, E. (1964) *Acta Phys. Hung.* **17**, 315   4.1, 5.5
Lendvay. E. (1965) *Magyar Fiz. Fol.* **8**, 231 (in Hungarian). English translation: *Air Force Cambridge Research Laboratory, Translation No. 51 (Ref. 69-0275; June 1965)*   3.5

Levy, F. and Mooser, E. (1972) *Helv. Phys. Acta* **45**, 902   A.1

Liesegang, R. E. (1896) *Naturwiss. Wochenschr.* **11**, 353   1.1, 5.1

Liesegang, R. E. (1897) *Z. phys. Chem.* **23**, 365   5.1

Liesegang, R. E. (1898) *Chemische Reaktionen in Gallerten,* Düsseldorf   1.1, 5.1

Liesegang, R. E. (1914) *Z. phys. Chem.* **88**, 1   4.3, 5.2

Liesegang, R. E. (1924) *Chemische Reaktionen in Gallerten* (new edition of 1898 work), Steinkopff Verlag, Dresden   1.1, 5.3

Liesegang, R. E. (1926) *Colloid Chemistry* (Editor, J. Alexander), Chemical Catalog Co, New York   1.1

Liesegang, R. E. (1937) *Kolloid Z.* **81**, 1   5.2

Liesegang, R. E. (1939) *Kolloid Z.* **87**, 57   5.2

Lifshitz, I. M. and Slezov, V. V. (1961) *J. Phys. Chem.* **19**, 35   5.4

Lloyd, D. J. (1926) *Colloid Chemistry* (Editor J. Alexander), Chemical Catalog Co, New York   1.2, 2.1, 2.2

Lloyd, F. E. and Moravek, V. (1928) *Plant Physio.* **3**, 101   5.2

Lovett, R., Ortoleva, P. and Ross, J. (1978) *J. Chem. Phys.* **69**, 947   5.4, 5.6

Lucchesi, P. J. (1956) *J. Colloid. Sci.* **11**, 113   5.3

Lüppo-Cramer, H. (1912) *Kolloid Z.* **13**, 35   1.1

Madelay, J. D. and Sing, K. S. W. (1962) *J. Appl. Chem.* **12**, 494   1.3, 2.1

Manegold, E. (1941) *Kolloid Z.* **96**, 186   2.2

Marqusee, J. A. and Ross, J. (1984) *J. Chem. Phys.* **80**, 536   5.4

Marriage, E. (1891) *Wiedemann's Ann.* **44**, 507   1.2

Matalon, R. and Packter, A. (1955) *J. Colloid Sci.* **10**, 46   3.1, 4.1

Mathur, P. B. (1961) *Bull. Chem. Soc. Jap.* **34**, 437   3.4

Mathur, P. B. and Ghosh, S. (1958) *Kolloid Z.* **159**, 143   5.6

Mazumdar, D., Govindacharyulu, P. A. and Bose, D. N. (1982) *J. Phys. Chem. Solids* **43**, 933   A.2

McCauley, J. W. (1965) M.S. Thesis in Geochemistry and Mineralogy, The Pennsylvania State University   1.4, 3.5

McConnell, J. D. C. (1960) *Min. Mag.* **32**, 535   3.5

McMasters, M. M., Abbott, J. E. and Peters, C. A. (1935) *J. Am. Chem. Soc.* **57**, 2504   5.6

Meal, L. L. and Meeks, F. R. (1968) *J. Colloid and Interface Sci.* **26**, 183   3.3, 5.4

Meeks, F. R. and Veguilla, L. A. (1961) *J. Colloid Sci.* **16**, 455   5.5

Miller, M. A. (1937a) *J. Phys. Chem.* **41**, 375   A.1

Miller, M. A. (1937b) *Kolloid Z.* **80**, 327   2.1, A.1

Mitchell, R. S. (1959) *Z. Krist.* **111**, 372   A.2

Miyamoto, S. (1937) *Kolloid Z.* **78**, 23   4.5, 5.2

Moeller, W. (1916) *Kolloid Z.* **19**, 209   5.2

Moeller, W. (1917) *Kolloid Z.* **20**, 242   5.2

Morse, H. W. (1930) *J. Phys. Chem.* **34**, 1554   5.4

Morse, H. W. and Pierce, G. W. (1903) *Z. phys. Chem.* **45**, 589   1.2, 2.3, 4.3, 5.1, 5.3, 5.4, 5.5

Morse, M. M. H. and Donnay, D. H. (1931) *Bull. Soc. Franc. Mineral.* **54**, 19   3.5

Mott, N. F. (1950) *Nature* **165**, 295   4.2

Müller, F. C., Kai, S. and Ross, J. (1982a) *J. Chem. Phys.* **86**, 4078   5.2

Müller, F. C., Kai, S. and Ross, J. (1982b) *Science* **216**, 635   5.2

Müller, K. K. (1939) Thesis. Technische Hochschule, Stuttgart   2.1

Mullin, J. W. (1961) *Crystallization,* Butterworths Scientific Publ., London   4.1

Murphy, J. C. and Bohandy, J. (1967) *Bull. Amer. Phys. Soc.* **12**, 327   1.3

Murray, L. A. (1964) *Electronic Ind.* **23**, 83   1.3

Nernst, W. (1904) *Z. phys. Chem.* **47**, 52   3.1

Nickl, J. and Henisch, H. K. (1969) *J. Electrochem. Soc.* **116**, 1258   1.3, 1.4, 3.5

Nicolau, I. F. (1980a) *J. Cryst. Growth* **48**, 45   1.3

Nicolau, I. F. (1980b) *J. Cryst. Growth* **48**, 51   1.3
Nicolau, I. F. and Joly, J. P. (1980) *J. Cryst. Growth* **48**, 61   1.3
Nutting, P. G. (1929) *J. Washington Acad. Sci.* **19**, 402   5.6
Obrist, J. (1937) *Kolloid Z.* **81**, 327   5.2
O'Connor, J. J., Dipietro, M. A., Armington, A. F. and Rubin, B. (1966) *Nature* **212**, 68   1.3
O'Connor, J. J., Thomasien, A. and Armington, J. J. (1968) *Air Force Cambridge Research Laboratories* (*Ref.* 68-0089) *Physical Sciences Research paper No.* 352   1.3
Orlovski, T. (1926) *Kolloid Z.* **39**, 48   5.2
Ortoleva, P. (1978) *Theoretical Chemistry IV* (Editor, H. Eyring), Academic Press, New York   5.4
Ortoleva, P. (1984) *Chemical Instabilities*, NATO-ASI Series L, Vol. 120, (Editors: G. Nicholis and F. Baras), Reidel, Dordrecht   5.3, 5.4, 5.5, 5.6
Ostwald, Wi. (1897a) *Z. phys. Chem.* **22**, 289   4.1
Ostwald, Wi. (1897b) *Z. phys. Chem.* **27**, 365   1.1, 3.2
Ostwald, Wo. (1925) *Kolloid Z.* **36**, 380   2.1, 5.6
Ostwald, Wo. (1926) *Kolloid Z.* **40**, 144   3.2
Ostwald, Wo. (1939) *Kolloid Z.* **89**, 105   5.1
Packter, A. (1955) *Nature* **175**, 556   5.3, 5.6
Packter, A. (1956a) *J. Colloid Sci.* **11**, 96   5.6
Packter, A. (1956b) *J. Colloid Sci.* **11**, 150   5.6
Palaniandavar, N., Gnanam, F. D. and Ramasamy, P. (1986) *Cryst. Res. and Technol.* **21**, 379   5.3
Palaniandavar, N., Kanniah, N., Gnanam, F. D. and Ramasamy, P. (1985) *Bull. Mat. Sci.* **7**, 105   5.4
Pandya, G. R., Dave, K. C. and Desai, C. F. (1986) *Cryst. Res. and Technol.* **21**, K30   A.1
Patel, A. R. and Arora, S. K. (1973) *J. Cryst. Growth* **18**, 199   1.3, A.1
Patel, A. R. and Rao, A. V. (1977) *J. Cryst. Growth* **38**, 288   A.1
Patel, A. R. and Rao, A. V. (1978) *J. Cryst. Growth* **43**, 351   4.4, A.1
Patel, A. R. and Rao, A. V. (1979) *Kristall und Tech.* **14**, 151   1.3
Patel, A. R. and Rao, A. V. (1980) *J. Cryst. Growth* **49**, 281   1.3
Perison, J. (1968) Personal communication, The Pennsylvania State University   2.1, 4.3
Pietro, J., García-Ruiz, J. M. and Amorós, J. L. (1980) *6th Int. Conf. on Cryst. Growth, Moscow, USSR,* September 10-16, 1980. Extended Abstracts, *Vol. IV,* 109   A.1
Pillai, K. M. and Ittyachen, M. A. (1978) *Ramāna (India)* **10**, 613   3.1, A.1
Pillai, K. M. and Ittyachen, M. A. (1979) *Current Sci.* **48**, 202   3.1
Pillai, K. M., Ittyachen, M. A. and Vaidyan, V. K. (1980) *Nat. Acad. Sci. Lett.* (*India*) **3**, 37   1.3
Pillai, K. M., Vaidyan, V. K. and Ittyachen, M. A. (1981) *Cryst. Res. and Technol.* **16**, K82   1.3
Pinsker, Z. G., Tatarinova, L. and Novikova, V. (1943) *Acta Physio-chim. USSR* **18**, 378   A.2
Plank, C. J. and Drake, L. C. (1974a) *J. Colloid Sci.* **2**, 399   2.1, 2.2, 4.4
Plank, C. J. and Drake, L. C. (1974b) *J. Colloid Sci.* **2**, 413   2.1, 2.2, 4.4
Popescu, C. and Henisch, H. K. (1975) *Phys. Rev. B.* **11**, 1563   3.1
Popp, K. (1925) *Kolloid Z.* **36**, 208   5.2
Prager, S. (1956) *J. Chem. Phys.* **25**, 279   5.3, 5.4
Prigogine, I. (1984) *Order Out of Chaos; Man's New Dialogue with Nature,* Bantam Books, New York   1.1
Prigogine, I. and Nicolis, G. (1967) *J. Chem. Phys.* **46**, 3542   5.4
Ramaiah, K. S. (1939) *Proc. Ind. Acad. Sci.,* **9A**, 467   5.3
Rayleigh, Lord (1919) *Phil. Mag.* **38**, 738   1.1
Reiss, H. (1950) *J. Chem. Phys.* **18**, 529   5.5
Ribet, M., Gits-Leon, S., Lefaucheux, F. and Robert, M. C. (1986) *Ferroelectrics* **66**, 259   A.2

Riegel, E. R. and Reinhard, M. C. (1927) *J. Phys. Chem.* **31**, 713   4.3

Ross, J. (1984) *Physica* **12D**, 303   5.6

Rothmund, V. (1907) *Löslichkeit und Löslichkeitbeeinflussung*, Leipzig   5.2

Roy, S. (1931) *Kolloid Z.* **54**, 190   1.3, 2.1, 4.5, 5.2, 5.4

Runge, G. (1855) *Der Bildungstrieb der Stoffe*, Oranienburg   1.1

Sahama, T. G. and Kanula, V. (1940) *Ann. Acad. Scient. Fennicae* **42A**, No. 3   2.1

Salvinien, J. and Moreau, J. J. (1960) *J. Chim. Phys.* **57**, 518   5.5

Sangwal, K. and Patel, R. (1974) *J. Cryst. Growth* **23**, 282   3.6

Saraf, A. G., Saraf, K. B., Wani, P. A. and Bhoraskar, S. V. (1986a) *Cryst. Res. and Technol.* **21**, 375   A.2

Saraf, A. G., Saraf, K. B., Wani, P. A. and Bhoraskar, S. V. (1986b) *Crys. Res. and Technol.* **21**, 449   A.1

Saunders, P. R. (1955) *Rheology of Elastomers*, Pergamon Press, London   2.1

Saunders, P. R. and Ward, A. G. (1955) *Nature* **176**, 26   2.1

Schleussner, C. A. (1922) *Kolloid Z.* **31**, 347   4.3, 5.4

Schleussner, C. A. (1924) *Kolloid Z.* **34**, 338   5.3

Schmidt, H. (1939) *Kolloid Z.* **89**, 45   5.2

Sears, G. W. (1961) *J. Phys. Chem.* **65**, 1738   4.1

Serb-Serbina, N. N. (1933) *Kolloid Z.* **62**, 79   5.3

Shanmugham, M., Gnanam, F. D. and Ramasamy, P. (1982) *Ind. J. of Pure and Applied Phys.* **20**, 579   A.2

Shanmugham, M., Gnanam, F. D. and Ramasamy, P. (1985) *Ind. J. of Pure and Applied Phys.* **23**, 82   A.2

Shanmugham, M., Gnanam, F. D. and Ramasamy, P. (1986) *J. Mat. Sci. Lett.* **5**, 174   A.2

Shewmon, P. G. (1963) *Diffusion in Solids*, McGraw-Hill, New York   3.1

Shinohara, S. (1970) *J. Phys. Soc. Jap.* **29**, 1073   5.4

Shinohara, S. (1974) *J. Phys. Soc. Jap.* **37**, 264   5.4

Shiojiri, M., Kaito, C., Saito, Y., Murakami, M. and Kawamoto, J. (1978) *J. Cryst. Growth* **43**, 61   A.1

Siebert, W. W. (1930) *Biochem. Z.* **220**, 487   5.6

Simon, A. L. (1913) *Kolloid Z.* **12**, 171   A.1

Smakula, A. (1962) *Einkristalle*, Springer Verlag, Berlin   4.1

Smoluchowski, V. (1916) *Phys. Z.* **17**, 557   5.4

Smoluchowski, V. (1917) *Z. phys. Chem.* **92**, 129   5.4

Spezie, G. (1899) *Atti. Accad. Torino* **34**, 705   1.1

Spotz, E. C. and Hirschfelder, J. O. (1951) *J. Chem. Phys.* **19**, 1215   3.2

Stansfield, R. (1917) *Amer. J. Sci.* **43**, 1   5.2

Stempell, G. (1929) *Biol. Zentralblatt* **49**, 10   5.6

Stern, K. H. (1954) *Chem. Rev.* **54**, 79   5.2

Stern, K. H. (1967) *Bibliography of Liesegang Rings*, National Bureau of Standards; Miscellaneous Publication No. 292   1.2, 5.2

Stonham, J. P. and Kragh, A. M. (1966) *J. Photog. Sci.* **14**, 97   2.1

Stranski, I. N. and Kaischew, R. (1939) *Sup. Fiz. Nauk.* **21**, 408   3.6

Sultan, F. S. A. (1952) *Phil. Mag.* **43**, 1099   3.1

Sunagawa, I. (1981) *Bull. Mineral.* **104**, 81   3.6

Suri, S. K. and Henisch, H. K. (1971) *Phys. Stat. Sol.* **44**, 627   A.2

Swanson, H. E. and Fuyat, R. K. (1953) National Bureau of Standards, Circular 534, II:51   3.5

Swyngedauw, J. (1939) *Kolloid Z.* **89**, 318   1.3

Tammann, G. (1922) *Aggregatzustände*, Leipzig. English translation by R. F. Mehl, published (1925) by V. Nostrand, New York   4.1

Thiele, H. and Micke, H. (1948) *Kolloid Z.* **111**, 73   2.1

Tillians, J. and Heublin, O. (1915) *Umschau* **19**, 930   5.2

Torgesen, J. L. and Peiser, H. S. (1968) *Methods and Apparatus for Growing Single Crystals of Slightly Soluble Substances*, U.S. patent 3 371 038 (February 27, 1968)   3.5

Torgesen, J. L. and Sober, A. J. (1965) *National Bureau of Standards Technical Note No.* 260, p. 13 (May)   1.3

Traube, J. and Takehara, K. (1924) *Kolloid Z.* **35**, 245   5.6

Treadwell, W. D. and Wieland, W. (1930) *Helv. Chim. Acta* **13**, 856   2.1

Trebbe, K. F. and Plewa, M. (1982) *Z. anorg. allgem. Chem.* **489**, 111   A.1

Turing, A. M. (1952) *Phil. Trans. Roy. Soc. B.* **273**, 37   5.4

Turnbull, D. and Vonnegut, B. (1952) *Indust. Eng. Chem.* **44**, 1292   4.1

Van Bemmelen, K. (1902) *Z. anorg. Chem.* **20**, 265 (there are many other relevant papers in the same journal, going back to 1878)   2.2

Vand, V. (1951) *Phil. Mag.* **42**, 1384   4.3

Vand, V. and Hanoka, J. I. (1967) *Mat. Res. Bull.* **2**, 241   4.3

Vand, V., Vedam, K. and Stein, R. (1966) *J. Appl. Phys.* **37**, 2551   3.1, 5.3

Van Hook, A. (1938a) *J. Phys. chem.* **42**, 1191   4.3, 5.5

Van Hook, A. (1938b) *J. Phys. Chem.* **42**, 1201   4.3

Van Hook, A. (1940) *J. Phys. Chem.* **44**, 751   5.3

Van Hook, A. (1941a) *J. Phys. Chem.* **45**, 422   5.6

Van Hook, A. (1941b) *J. Phys. Chem.* **45**, 1194   5.6

Van Hook, A. (1963) *Crystallization: Theory and Practice*, Reinhold, New York   4.1

Van Oss, C. J. and Hirsch-Ayalon, P. (1959) *Science* **129**, 1365   5.3

Van Rosmalen, G. M., Marchée, W. G. J. and Bennema, P. (1976) *J. Cryst. Growth* **35**, 169   A.1

Vasudevan, S., Nagalingham, S., Dhanasekaran, R. and Ramasamy, P. (1981) *Cryst. Res. and Technol.* **16**, 293   3.1

Veil, S. (1935) *Les Periodicités de Structure*, Paris   5.4

Venzl, G. and Ross, J. (1982) *J. Chem. Phys.* **77**, 1302   5.5

Vincent, B., Bijsterbisch, B. H. and Lyklema, J. (1971) *J. Colloid and Interface Sci.* **37**, 171   5.4

Volmer, M. (1932) *Trans. Faraday Soc.* **28**, 359   3.4

Volmer, M. (1939) *Kinetik der Phasenbildung*, T. Steinkopff Verlag, Dresden   3.1

Volmer, M. and Weber, A. (1926) *Z. phys. Chem.* **119**, 225   4.1

Vonnegut, B. (1947) *J. Appl. Phys.* **18**, 593   4.1

Vonnegut, B. (1949) *Chem. Rev.* **44**, 277   4.1

Wagner, C. (1950) *J. Colloid Sci.* **5**, 85   5.4, 5.5

Wagner, C. (1961) *Berichte der Bunsengesellshaft Phys. Chem.* **65**, 581   5.4

Wakim, F. G., Henisch, H. K. and Atwater, H. A. (1965) *J. Chem. Phys.* **42**, 2619   A.2

Ward, A. G. (1954) *Brit. J. Appl. Phys.* **5**, 85   2.1

Warren, B. E. (1940) *Chem. Rev.* **26**, 237   2.2

Washburn, J. (1958) *Growth and Perfection of Crystals* (Editors, R. H. Dorman B. W. Roberts and D. Turnbull), John Wiley & Sons, New York   3.1

Weimarn, P. P. von (1926) *Colloid Chemistry* (Editor, J. Alexander), Chemical Catalog Co, New York   1.2

Williams, D. J. A. and Ottewill, R. H. (1971) *Kolloid Z.* **243**, 141   5.4

Zettlemoyer, A. C. (1969) *Nucleation*, Dekker, New York   5.5

INDEX